The Secret Lives of Numbers

The Secret Lives of Numbers

KATE KITAGAWA

AND

TIMOTHY REVELL

VIKING

an imprint of

PENGUIN BOOKS

VIKING

UK | USA | Canada | Ireland | Australia
India | New Zealand | South Africa

Viking is part of the Penguin Random House group of companies
whose addresses can be found at global.penguinrandomhouse.com

First published 2023
001

Copyright © Kate Kitagawa and Timothy Revell, 2023

The moral right of the authors has been asserted

Set in 12/14.75pt Bembo Book MT Pro
Typeset by Jouve (UK), Milton Keynes
Printed and bound in Great Britain by Clays Ltd, Elcograf S.p.A.

The authorized representative in the EEA is Penguin Random House Ireland,
Morrison Chambers, 32 Nassau Street, Dublin D02 YH68

A CIP catalogue record for this book is available from the British Library

HARDBACK ISBN: 978–0–241–54411–2
TRADE PAPERBACK ISBN: 978–0–241–54412–9

Contents

List of Illustrations

The author and publisher gratefully acknowledge the permission granted to re-produce the copyright material in this book. Every effort has been made to trace copyright holders and to obtain their permission. The publisher apologises for any errors or omissions and, if notified of any corrections, will make suitable acknowledgment in future reprints or editions of this book.

Prelude

In a scene in the American political drama *The West Wing*, two senior government aides stare in disbelief at a slide in a presentation. A group of cartographers is attempting to explain that the map of the world, the map they have known and trusted all their lives, is just one of many. And it is flawed. 'Are you saying the map is wrong?' asks one aide incredulously.

No map of the planet is accurate: it's simply not mathematically possible. The surface of a sphere cannot be transformed into a two-dimensional drawing without distortion. But, as the cartographers explain, the map they are looking at promotes a Eurocentric view of the world. Europe appears to be larger than South America, but South America is actually twice as large. Germany is located in the middle of the map when it's actually in the northernmost quarter of Earth. All this time, our view of the world has been distorted.

The map was drawn by Flemish cartographer Gerardus Mercator in the sixteenth century. It was originally intended for sailors crossing oceans, not for policy wonks considering geopolitics. It has been passed down from generation to generation, solidifying its place as the dominant map, giving people across the planet the impression that this is what the world looks like, not that it is just one perspective.

The history of mathematics is similar. Despite mathematics' reputation as the study of fundamental truths, cold hard calculations and irrefutable proofs, it has not escaped the powerful individuals and structures that have shaped truth and knowledge. Far from it: in fact, the history of mathematics has accumulated biases over thousands of years – from the way certain mathematics and mathematicians are revered to the stories we tell about its origins. It is time this pattern was re-examined and the story retold.

When we, Kate and Timothy, first met to discuss a joint book project, we hadn't anticipated where it would lead. Sipping tea in a

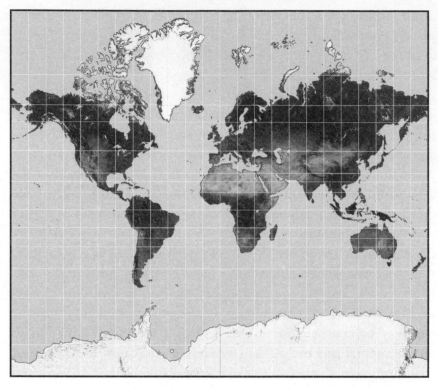

The Mercator projection.

bookshop in Charing Cross, London, we talked about our shared love of mathematics and agreed that we should write an approachable history of the subject. Drawing on Kate's expertise as a mathematical historian and Timothy's mathematical and journalistic credentials, we thought it would be straightforward.

We were wrong. The more we delved into the history of mathematics, the more we uncovered about the way it has been distorted. And the more we felt compelled to do something about it.

The origins of mathematics are beautifully varied. Rather than ideas springing up in one location, variations have often appeared throughout history, demonstrating just how powerful the human propensity to reason really is. Ideas ignore national borders, so often mathematics has spread from one place to another alongside trade and cultural exchange. However, the progress of mathematics is not linear. It has gone forwards and backwards, jumped around the planet, gone off on tangents, on adventures and sometimes down dead ends. And it is much the richer for it. Despite its reputation for logical progressions, mathematics is a far more chaotic affair.

However, this is not how the story of mathematics is usually told. The ancient Greeks are put on a pedestal as somehow being the originators of modern mathematics, yet so much of what is now incorporated into our global knowledge comes from many other places too, including ancient China, India and the Arabian peninsula. This assumption that the European way of doing things is superior didn't originate in mathematics – it came from centuries of Western imperialism – but it has infiltrated it. Mathematics outside of ancient Greece has often been put to one side as 'ethnomathematics', as if it were a separate subject altogether, a side story to the real history.

As we worked our way through thousands of years of mathematics, almost everything we thought we knew was challenged in one way or another. Some well-known stories ended up being misrepresentations and others complete fabrications. Many mathematicians and mathematics have wrongly been excluded from history. Over the following pages, we will reveal some of the ways in which the story of mathematics has been warped. The real story is one of a truly global endeavour. Mathematics is about ideas and inventing ways to think them through to their conclusions. Diversity of thought isn't just important in mathematics, it is fundamental.

Take calculus. This mathematical theory for describing and determining how things change over time is one of the most important and useful advances in human history. It is crucial to engineering – without it, we couldn't accurately build bridges or rockets – and it is used in nearly every scientific discipline to help us better understand the world. Many aspects of our lives today would not be possible without it.

So who takes the credit? The usual story is that Isaac Newton, an English mathematician, and Gottfried Wilhelm Leibniz, a German one, independently developed their own versions of calculus at around the same time in the seventeenth century. This much is true, but to know only this is akin to looking at Mercator's map – it is a distorted view. There is a far earlier claim to the ideas behind calculus.

In the fourteenth century, a school in Kerala, India, was a melting pot for mathematicians. Its founder, Mādhava of Sangamagrama, was a brilliant mathematician, and among his achievements is describing a theory of calculus. He explored the key ideas that make calculus possible, which were then honed by successive mathematicians at the Kerala school. This theory was neither complete nor perfect, but that's always the case with something new. Many of the first light bulbs burned out too quickly and the glass turned black because of flaws in their design, but Thomas Edison is still recognized for his role in that nineteenth-century invention. It's time we recognized Mādhava too.

Ideas are at the forefront of any history of mathematics, but they cannot be separated from the people who had them. To truly represent the origins of mathematics we must look at the origins of mathematicians too. Some of those featured in this book were not just impressive mathematicians but also broke down barriers to help make mathematics a more inclusive and global subject. In this book, we give greater prominence to these forgotten mathematicians and explain how they fit into the traditional story, as well as correcting mistruths and misrepresentations about them. Important people that didn't fit the accepted idea of a mathematician were not just oppressed during their lifetimes but have faced a continued attack from historians and commentators since.

Take Sophie Kowalevski,* who was born in Moscow in 1850, just before the Crimean War. During her life, she was constantly discouraged and forbidden from pursuing mathematics. Her father refused her access to a proper education, believing that having a daughter who was a learned woman would bring shame upon him. Views like this were common at the time. Despite this, she pursued mathematics and produced work that was easily good enough to earn a doctorate.

---

* Kowalevski's name has been written in many different ways. She tended to use 'Sophie Kowalevski' in her academic publications so we have chosen to use that version too.

Yet because of her gender many universities wouldn't let her take the exam to obtain one.

Through immense determination, Kowalevski eventually managed to earn a position at Stockholm University, becoming the world's first female mathematics professor. Even then, her professorship was unpaid – she had to personally collect money from her students in order to survive. Some people were unhappy that she had even attained such a position. The famed playwright August Strindberg described the concept of a female professor as a 'pernicious and unpleasant phenomenon'.[1]

After her death, Kowalevski's legacy was distorted by some biographers who too often relied on gender stereotypes to tell her story rather than the facts of her life. She was an extraordinary mathematician yet was presented as a sort of femme fatale who relied on her looks and charm to make progress, despite there being little evidence for this. It's time to put an end to the tainting of stories like Kowalevski's.

We believe that this retelling of the story of mathematics is important, but we hope that it will be more than that. Mathematics has been filled with fascinating characters for millennia. It is a subject in pursuit of truth, of eye-opening ways of thinking and theorems to blow your mind. It is not a passionless pursuit but a creative one. As Kowalevski once said, 'It is a science which requires a great amount of imagination.'[2] The history of mathematics is an unmissable saga of the highest calibre.

No single book can right every wrong or tell a truly complete history, but in the same way that a new map can change how we view the world, a new history can do the same. In ours, we tell the story of mathematics as it really is – beautifully chaotic and collaborative. Mathematics today is an awe-inspiring amalgamation of concepts from all over the world pioneered by a group of mathematical boundary-smashers, people who ignored the limitations society placed on them because of their race, gender and nationality. Mathematics is a subject with a rich and diverse history. It is time to tell it.

1. In the Beginning

Our species, *Homo sapiens*, has been around for 300,000 years but, as far as we can tell, mathematics is a relatively recent invention. Many artefacts have been lost or simply haven't lasted, so we have only a partial picture. The first traces of human mathematical activity start to appear around 20,000 years ago, in the form of scratched tally marks on animal bones.

One of the oldest and most famous of these is the Ishango bone, which was found along the border between modern-day Uganda and the Democratic Republic of the Congo and dates from 20,000–18,000 BCE. The bone is probably the fibula of a baboon, though it could be from a wolf or similar-sized animal. It has a piece of quartz attached to the top, suggesting it may have been used as a tool. Running down its length are three columns populated with tally marks. The scratches may be there simply for gripping the tool, but there may also be more to them than that.

Ishango bones (*front and back*).

The marks on the first column add up to 48, and the marks on the second and third columns add up to 60. Each of the columns is split into distinct segments, with the third column's split the most interesting. The sixty notches there are split into groups of 11, 13, 17 and 19. These are prime numbers – numbers that can only be divided by 1 and themselves. Prime numbers are undoubtedly some of the most important numbers in mathematics. As later mathematicians will discover, they are the building blocks of all other numbers. To see them here, on a carving from more than 20 millennia ago, is like receiving a message from an alien. It is exhilarating and surprising – yet it is also hard to know exactly what it means.

The mathematical patterns could just be coincidence, but they could also show the numerical sophistication of our ancient ancestors. The numbers 48 and 60 are 4×12 and 5×12 respectively, hinting that the people who made the scratches had a number system built around the number 12 (rather than 10, as we use today). One of the earliest number systems we know of was built around the number 60, so this is far from implausible. Another option is that the bone was a six-month lunar calendar, with the notches representing phases of the moon. Another, put forward by twentieth-century mathematician Claudia Zaslavsky, was that the bone may have been used by a woman to track her menstrual cycle. Measuring the coming and going of the seasons for planting seeds, or when rivers would flood, also seems like a reasonable possibility. Similar bones have been discovered in other parts of Africa and elsewhere. For tens of thousands of years, it appears that counting has been an integral part of being human.

The very earliest surviving signs of mathematics are much like the Ishango bones. Such relics may show a giant conceptual leap for our species, a moment when we began to think in the mathematical abstract – or they may simply be scratches. Remains of ancient monuments and pottery often feature elaborate geometric designs, but does that mean the makers understood the mathematics behind the displays, or did they just like the patterns?

The earliest mathematics our species developed may not have been written down or have left any physical trace. More contemporary evidence shows that a deep understanding of mathematics can develop through speech alone. The Akan people of West Africa, for example, had a sophisticated set of mathematical tools for dealing with weights and measures that was passed down by word of mouth. The oral nature of their mathematical system made it perfect for doing business with Arab and European merchants between the fifteenth and late nineteenth centuries. However, it also meant that it was demolished by hundreds of years of the Atlantic slave trade. After researchers managed to reconstruct how it worked in 2019, using the few remaining artefacts held in museums, they suggested the system should be given UNESCO World Heritage status because it was so spectacular.

In this case, the system was in use until fairly recently and some artefacts still exist, but there have probably been many other oral mathematical systems that are now lost to time. Counting, and its consequences, was probably integral to many communities and civilizations that never had the need to write anything of this nature down. Or if they did, any trace of it has now perished. These first moments of mathematics are vague and will remain so for ever. However, with the advent of written language and the rise of some of the world's biggest civilizations, the picture becomes a little clearer.

By the rivers of Babylon

Between the Tigris and Euphrates rivers is a stretch of fertile land that has been home to many great ancient civilizations. The rivers have their sources separately in what is now Turkey, meandering through modern-day Iraq, Syria and Iran before flowing into the Persian Gulf. Together they form a natural border for the area that was once known as Mesopotamia.

By 3000 BCE, the Sumerian civilization was thriving here. The Sumerians built complex cities with vast irrigation systems. They also had one of the earliest legal systems, complete with courts, jails and government records. They had developed the earliest known writing system, cuneiform – needed for those records – and a counting system, to boot. They even created a postal service.

Over the next thousand years, the Akkadians became the dominant force in the region. They brought with them their own technology, including the abacus – a tool they had invented. (It worked slightly differently to later versions, such as the Chinese one.) Eventually, their empire would fall, leaving behind two distinct Akkadian-speaking groups: Assyrians in the north and Babylonians in the south. Each produced a vast civilization of its own, but it was in the south that mathematics really accelerated.

The city of Babylon, roughly 100 kilometres south of modern-day Baghdad, was the capital of the empire of Babylonia. Under the direction of King Hammurabi, who ruled from around 1792 to 1750 BCE, Babylonia became a force to be reckoned with. He controlled several city-states in the region, making Babylonia extremely rich and powerful.

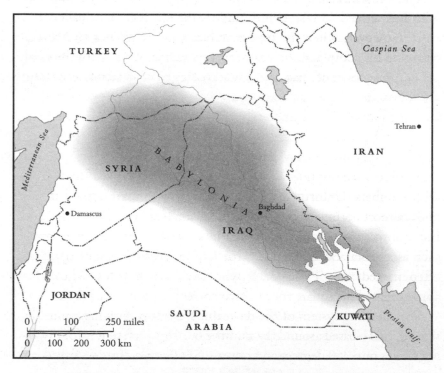

Mesopotamia covered parts of current-day Iraq and Syria.
Babylonia was an Akkadian-speaking state, located in the
central-southern part of Mesopotamia, and its main city was Babylon.

This provided the stability and resources needed for a mathematical community to develop and thrive.

An extensive collection of clay tablets that still survive today record many details about Babylonia in this era. Scribes would scratch what they wanted to record with a sharpened stick on wet clay and then leave it to harden in the sun. These tablets, to Babylonians, were what paper and spreadsheets are to us today – crucial tools for record-keeping. They recorded Hammurabi's legal system, known as the Code of Hammurabi, which consisted of 282 written laws and contained one of the earliest examples of being innocent until proven guilty – though how guilty you were depended on whether you were a person with property, free or a slave. They also recorded transactions and told stories, including myths about creation, and relayed news.

One tablet has survived that is essentially a bad review. Written around 1750 BCE, it is from an unsatisfied customer called Nanni, who had agreed to buy copper ingots from a merchant called Ea-nasir. However, when the ingots arrived, they were not to Nanni's liking. In his complaint, he wrote that he was unhappy with the copper and that the seller had been rude to his servant when completing the transaction. Scraping and baking a review into a form that would last for thousands of years displays consumer power at its finest.*

The Babylonians used mathematics for many practical purposes, including splitting plots of land and calculating tax. Some clay-tablet writers recorded revenues and budgets, and so familiarized themselves with numbers. Unfortunately, they did not sign their names, so we know almost nothing about individual mathematicians from this time. But some certainly studied mathematics systematically, taking in topics such as algebra and uncovering that famous theorem about triangles often named after Pythagoras (who lived much later). They also approximated the square root of two correct to six decimal digits.

The counting system of the day came from the Sumerians and was sexagesimal – based around the number 60. Our preferences for dividing circles into 360 degrees and hours into 60 minutes stems from this

---

* And, of course, if you wanted to use that consumer power to bake a favourable review of this book, we and future archaeologists would be most grateful.

system. Below are the cuneiform symbols they used to represent the numbers 1 to 59:

1	2	3	4	5	6	7	8	9	10
𒁹	𒈫	𒐈	𒐉	𒐊	𒐋	𒐌	𒐍	𒐎	𒌋
11	12	13	14	15	16	17	18	19	20
𒌋𒁹	𒌋𒈫	𒌋𒐈	𒌋𒐉	𒌋𒐊	𒌋𒐋	𒌋𒐌	𒌋𒐍	𒌋𒐎	𒎙
21	22	23	24	25	26	27	28	29	30
𒎙𒁹	𒎙𒈫	𒎙𒐈	𒎙𒐉	𒎙𒐊	𒎙𒐋	𒎙𒐌	𒎙𒐍	𒎙𒐎	𒌍
31	32	33	34	35	36	37	38	39	40
𒌍𒁹	𒌍𒈫	𒌍𒐈	𒌍𒐉	𒌍𒐊	𒌍𒐋	𒌍𒐌	𒌍𒐍	𒌍𒐎	𒐏
41	42	43	44	45	46	47	48	49	50
𒐏𒁹	𒐏𒈫	𒐏𒐈	𒐏𒐉	𒐏𒐊	𒐏𒐋	𒐏𒐌	𒐏𒐍	𒐏𒐎	𒐐
51	52	53	54	55	56	57	58	59	
𒐐𒁹	𒐐𒈫	𒐐𒐈	𒐐𒐉	𒐐𒐊	𒐐𒐋	𒐐𒐌	𒐐𒐍	𒐐𒐎	

Babylonian cuneiform numerals.

The Babylonian number system was a positional system like ours, meaning that the order the numbers are written in tells you something about the amounts they represent. For example, when we write the number 271, it is with the implicit understanding that the number furthest to the right represents one unit, then, moving towards the left, there are seven tens and two hundreds. Or, in numbers

$$271 = \left(2 \times 10^2\right) + \left(7 \times 10^1\right) + \left(1 \times 10^0\right)$$

Similarly, the Babylonians used positions to represent powers of 60, so 271 could be expressed as

$$271 = \left(4 \times 60^1\right) \times \left(31 \times 60^0\right)$$

Or, in cuneiform

Where the Babylonian number system differs most from ours is that it had no zero — a true zero would not arise until much later. This meant that Babylonians would often have to work out the size of a number from context. If they saw the cuneiform symbol for 42, for instance, they would have to infer whether that meant 42, or 42×60^1, or 42×60^2, or $\dfrac{42}{60^1}$, or $\dfrac{42}{60^2}$, to name just a few of the options. Though this did sometimes lead to mistakes, it's not as unreasonable as it may first seem. If you heard someone say that a house costs '300' of a particular currency, depending on where in the world you are you could probably work out if it meant 300; 300,000; 3 million; or more.

Base-60 may initially seem complicated compared to base-10, but it gave the Babylonians a mathematical edge. The number 60 is a superior highly composite number, meaning that it has many factors — it can be divided by 1, 2, 3, 4, 5, 6, 10, 12, 15, 20, 30 and 60. This makes it easy to work with, particularly when writing fractions.

Recall that just as positions going left from a decimal point represent units, tens, hundreds, and so on, when going right after the decimal point they represent tenths, hundredths, thousandths, and so on. The number 0.347, say, is really shorthand for

$$0.347 = \frac{0}{10^0} + \frac{3}{10^1} + \frac{4}{10^2} + \frac{7}{10^3}$$

Now take the fraction $\dfrac{1}{3}$. In decimal, this is written as

$$0.333\ldots = \frac{0}{10^0} + \frac{3}{10^1} + \frac{3}{10^2} + \frac{3}{10^3} \cdots$$

We're so used to writing a third like this in decimal that its recurring nature seems normal, but it is a quirk of our number system. It comes from the fact that 10 cannot be divided by 3. However, 60 can. A third is the same as $\dfrac{20}{60}$, meaning that, in sexagesimal, it could simply be written as .20 or, in other numbers:

$$\frac{1}{3} = \frac{0}{60^0} + \frac{20}{60^1}$$

As 60 is a superior highly composite number, there are more fractions that can be expressed nicely in base-60 than in base-10.

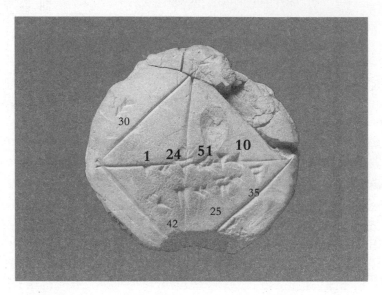

Babylonian approximation for $\sqrt{2}$.
In sexagesimal, 1 24 51 10. In decimal, approximately 1.414213.

The ancient Egyptians made similar advances around this time. From around 3000 BCE, the people there had specific symbols to represent different numbers as part of a base-10 system. A single line represented the number 1, two lines the number 2, and so on, up to the number 9. There were then specific hieroglyphs for numbers such as 10, 100, 1,000, and so on, as well as symbols for fractions. To write a given number, ancient Egyptians simply listed the correct combination of hieroglyphs.

Much of this is collated in the Rhind* papyrus, a manuscript

---

* The name comes from British archaeologist Alexander Henry Rhind, who purchased the papyrus in 1863. Most of it is now housed at the British Museum in London.

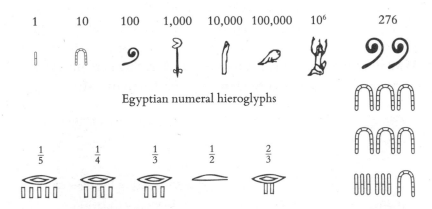

Egyptian numeral hieroglyphs

Hieroglyphics for some fractions

written by a scribe called Ahmes. It is the oldest surviving mathematics textbook we know of and has this extraordinary opening: 'Accurate reckoning. The entrance into the knowledge of all existing things and all obscure secrets.'[1] Ahmes wrote the manuscript in around 1550 BCE and says that he used texts from around 2000 BCE to put it together. That the mathematics it contains could be at least four thousand years old is hard to fully appreciate, especially considering that so much of what it contains resembles mathematics as we know it today.

The textbook contains eighty-four mathematical problems and ways to solve them. Six of the problems are about calculating the slope of a pyramid from its height and width using ideas akin to trigonometry. Mathematics is shaped by the people who develop it, so it is no surprise that Egyptian mathematicians were interested in the mathematics of pyramids when the pharaohs were so obsessed with building them. But mathematical ideas are also universal. Many other cultures independently uncovered the mathematics of trigonometry, from ancient China to Renaissance Europe, just with different motivations. The papyrus also includes division and multiplication tables, as well as explanations of how to calculate volume and area. Many of our modern-day concepts and ideas around arithmetic, algebra and geometry appear in one form or another.

There is some overlap between the ideas in the Rhind papyrus and those that appear on Babylonian tablets. The two civilizations had different number systems, beliefs and cultures, yet each uncovered similar mathematical truths. It is generally thought that this wasn't due to any active exchange but merely down to them independently exploring some of the most fundamental mathematical ideas.

A scroll under the arm

Across the Atlantic at around the same time another civilization was developing a different take on mathematics, born out of astronomy. The Maya civilization began around 2600 BCE. It wasn't a single empire but rather a collection of independent rulers who controlled city-states spanning from present-day Mexico and Honduras that shared a common culture, mythology and calendar. There were temples aligned to the movements of the sun, the moon and the planets, and cities that were vast and sprawling. Tikal, in what is now northern Guatemala, is thought to have had around fifty thousand inhabitants and three thousand separate buildings, ranging from palaces and shrines to houses, plazas and water reservoirs. It was an economic and ceremonial hub, with an extensive trade in precious commodities such as jade, quetzal feathers and cacao. The Maya, like other civilizations, had sophisticated irrigation systems for feeding their crops. They also developed a system of purifying water for drinking using zeolite minerals that is still in use today.

Mathematicians were important and famous enough to appear in wall paintings, their scrolls pictured under their arms. These mathematicians were aided by certain fortuitous features of Maya culture. For one, even though the Maya spoke many different local languages, there was only one writing system, consisting of hieroglyphs for syllables and the profiles of gods for numbers (pictured overleaf). Most people were illiterate, but scribes could communicate regardless of the language they spoke through books written in hieroglyphic script on paper made from the inner bark of fig trees.

The Maya also had another number system that was less decorative and more practical. It used two symbols: a dot and a bar. The dot

The Maya civilization. The area covers parts of today's Mexico, Guatemala, Belize, Honduras and El Salvador.

represented 1 and the bar 5. Rather than being built around 10 or 60, like decimal and sexagesimal number systems, the Maya number system was vigesimal, meaning that it was built around the number 20.

Unfortunately, understanding the exact workings of this number system involves a bit of guesswork. When the Spanish conquistadors invaded Mesoamerica in the sixteenth century, there were many Maya fig-bark books still in existence, but Catholic priests believed they contained the 'lies of the devil' and so had many of them burned.

Despite this, we know the Maya used their number systems to great effect. One of the central roles for Maya mathematicians was to be astronomers. Their job was to help plan sacred rituals so that they aligned with celestial events. The Maya built simple yet functional observatories to help predict the changing seasons and when it would be best to plant crops. Though the building itself was constructed later, many of the windows of the Caracol observatory in Chichén

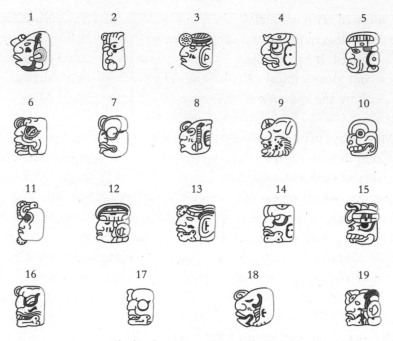

The head-variant numeral glyphs.

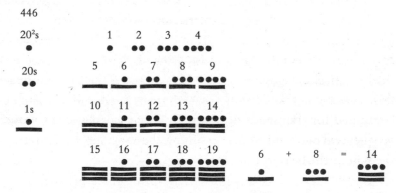

Maya numerals and a sample addition (6 + 8 = 14).

Itza in what is now Mexico gave the perfect view of important astronomical events, such as the setting of the sun on the spring equinox.

The remains of a writing room of Maya scribes, known as the house of calendars, show us how astronomers kept track of their data. Dating from the early ninth century BCE, the wall and ceiling are adorned with colourful paintings, including several human figures, as well as numbers and glyphs. The wall was probably used like a blackboard. It has hieroglyphs decorated in colour, used for calendrical and astronomical calculations. The traces of two calculation tables show the movement of the moon, and possibly of Mars and Venus.

Most Maya mathematician-astronomers were from the priest class and held in high regard. They were able to accurately predict solar eclipses and even managed to predict the strange movements of Venus in the sky, which repeat over an eight-year period, partially due to the sun blocking our view of it. They viewed Venus as a companion of the sun, giving it the name Chak Ek', or Great Star.

The Maya made incredibly accurate measurements of the movements of the moon and stars: for instance, they calculated that 149 lunar months lasted 4,400 days; in our notation, this results in a lunar month of 29.5302 days, and today, we have it as 29.5306. Similarly, they worked out the length of the year as 365.242 days; today, we put this at 365.242198 days.

Driven by a desire to better understand the night sky and its effects on Earth, the Maya were pushed to develop mathematics. They believed that by mastering astronomy they would be successful in agriculture. A similar mechanism would drive another period of mathematical development elsewhere in the world, which began at least as early as the development of mathematics in Babylonia and continued for thousands of years. In China, mathematics would go way beyond rains and crops, becoming about the authority to rule and the will of the heavens.

2.   The Turtle and the Emperor

The legend goes that one day, around four thousand years ago, Yu the Great was taking a break from his duties as Emperor of China to walk along the banks of the Yellow River. As he gazed across the flowing water, he felt a dark moving object at his feet. He looked down and saw that it was a turtle. But it wasn't just any old turtle. He peered more closely. The turtle's shell had cracks in it that formed a three-by-three grid of Chinese numerals that he quickly recognized. It was a symbol of mathematical perfection.

The pattern he saw is now known as a magic square, and can be transcribed like this:

4	9	2
3	5	7
8	1	6

Notice how each column, row and diagonal adds up to fifteen. In the eyes of the people of ancient China, this numeric coincidence was an auspicious sign. Emperors were the most important figures of the state and performed rites in order to ensure that the harmony of the cosmos was preserved. Yu founded China's oldest dynasty, the Xia, and had full responsibility for what happened throughout its empire, with the outcomes of divination playing an important role in everything from battle to childbirth, illness and the harvest. On finding this auspicious pattern, Yu obtained the authority to present himself as the righteous leader of the land. He had the so-called Mandate of Heaven.

It is with stories like this that mathematics begins in China. Over the course of a thousand years, mathematics and divination were at the heart of each dynasty. Rulers relied on it both for practical purposes such as trade and for divine guidance, using mathematical methods to attempt to find out what the universe had in store for them. Mathematics in ancient China was power.

Though often under-appreciated outside of East Asia, the mathematics developed during this era was sophisticated, elegant and way ahead of its time. Magic squares, for example, first appeared in China but eventually cropped up in India, the Middle East and, much later, in Europe. This would become a pattern. Throughout history, mathematicians across the world would hit upon what they believed to be new discoveries, and only later would it be realized that these discoveries had been made in China hundreds – if not thousands – of years before.

Counting rods (and blessings)

The oldest records we have of mathematics in China are bones – specifically, bones used for divination. Even though Yu had stumbled upon a message-carrying turtle by chance, it was common for diviners to try to force the issue. They would try to speak directly to the gods by scratching questions into the shells of dead turtles or on the shoulder bones of cattle, which they would then heat until they

A replica of a turtle shell used in divination.

cracked. The resulting patterns would be interpreted as celestial answers to the questions they posed.

Yu's life pre-dates the oldest written records in China by hundreds of years, so the facts of his life are not universally agreed upon. What we know about him comes from stories that were passed down orally for generations and only recorded much later. However, there are many remains of these divinations on so-called oracle bones. And it is here that we can also see the oldest Chinese number system in action as part of oracle-bone script – an early ancestor to modern Chinese characters from around the fourteenth century BCE.

Oracle-bone script was a base-10 number system, but it was not positional. Instead, numerals were combined to represent bigger numbers (see the third row below); however, this had its limitations. The largest number archaeologists have found on an oracle bone is 30,000.

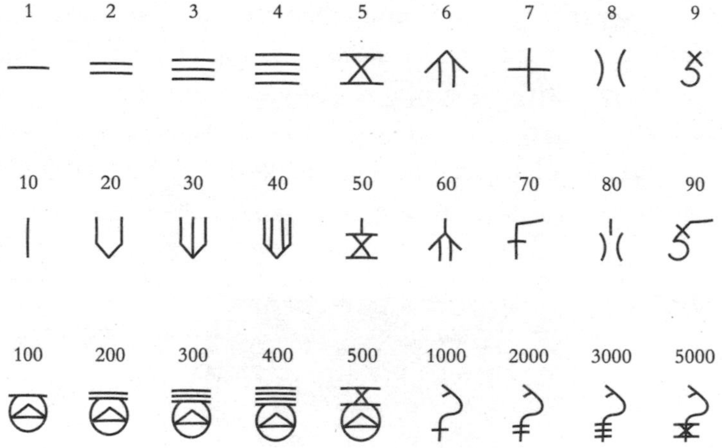

Oracle-bone numerals.

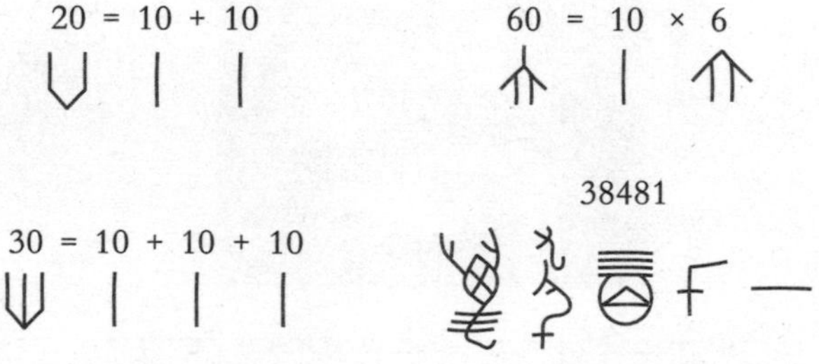

20 = 10 + 10   60 = 10 × 6

38481

30 = 10 + 10 + 10

Oracle-bone numerals could also be used for basic fractions. This can be seen in the oldest known decimal multiplication table, found when nearly 2,500 bamboo strips from around 300 BCE were donated to Tsinghua University in Beijing in 2008. It is not known exactly where the strips originally came from, but before reaching the university they were probably put up for sale after an illegal excavation.[1] Among the collection were twenty-one bamboo strips that form a multiplication table showing how to multiply any whole or half number between 0.5 and 99.5. These sorts of tables were used like calculators to quickly compute complicated sums. The ancient Babylonians had multiplication tables some four thousand years ago, earlier than the Chinese tables, though they were not in decimal. The earliest known European multiplication tables date from the Renaissance.

*c.*300 BCE bamboo strips.

Sometime after these bamboo strips were made, a different number system emerged in China that would prove particularly useful for traders – rod numerals. Rod numerals used line-based symbols that could easily be scratched into mud or sand, though many people used actual physical rods. Merchants in China at this time were part of the wealthy land-owning elite and would carry around with them a bundle of bamboo counting rods for performing on-the-fly calculations.

At the heart of the rod numeral system was a cunning trick for building up bigger numbers from smaller ones. Two systems to represent each of the numbers from 1 to 9 were used simultaneously. In the first, rods were placed vertically to indicate the numbers 1 to 5, then, in the numbers 6 to 9, one horizontal rod represented 5, and vertical rods were added for each successive number. In the second, the horizontal and vertical rods played the opposite role. For example, looking at the table below, you can see how 7 can be written as either two vertical rods and a horizontal one or two horizontal ones and a vertical one.

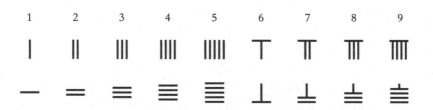

Examples of counting rods representing numbers.

To represent higher numbers, these two different systems for the numbers 1 to 9 were used alternately, with the predominantly vertical rods used to denote units, the predominantly horizontal rods indicating tens, and so on. The number 264 could be written as:

Even though there was no notation to mark a zero, it was possible to imply it by using one of the systems (vertical or horizontal) in consecutive order. For example, the number 209 would be written:

$$\| \qquad \text{ⅢⅢ}$$

The absence of a horizontal number between the two hundreds and the nine units makes it clear there are 0 tens. Hundreds of years before anyone came up with a symbol for zero (a story that we will return to in Chapter 5), Chinese mathematicians understood its utility as a placeholder. This idea of numbers being built up of digits in particular positions – as was the case in both ancient China and Babylonia – was truly revolutionary. It's hard to imagine now, when we are so familiar with using a position-based numerical system, but this leap was the mathematical equivalent of inventing the jet engine after previously relying on simply flapping your arms and hoping to take flight. By switching between the two methods to express digits, ancient merchants and mathematicians could zoom down the number line to higher numbers without having to invent new symbols or names for them – a small set of symbols and their positions were all you needed to know.

Counting rods may have been invented in China, although there is some evidence that they came from India. Either way, they took off in China and were a boon to the people who used them. By learning simple algorithms involving moving physical rods, traders could perform addition, subtraction, multiplication and division quickly and easily. To multiply two numbers, rods were laid on to a surface and combined in each position. There were even methods for using such rod manipulations to find square roots or solve simultaneous equations – equations involving more than one unknown quantity. The ancient Chinese also understood negative numbers, using black rods to represent positive numbers and red ones for negative numbers – though negative numbers never appeared in answers, only in calculations. Negative numbers may seem natural today, but throughout much of history numbers were so closely linked to physical objects that many mathematical civilizations outside China simply didn't consider the possibility that negative ones could be useful. 'Minus seven sheep' just didn't seem to make much sense. As we shall see in the next section, Chinese mathematics was greatly influenced by having opposites in balance, so one possibility is that this viewpoint helped them to more easily accept the idea of negatives.

The counting-rod system was an incredible innovation. For centuries, it would remain an integral part of Chinese calculation and trade until eventually superseded by the supercomputers of the time: abacuses. These were easier and faster to use than counting rods and became the dominant numerical tool in China by around 190 BCE.

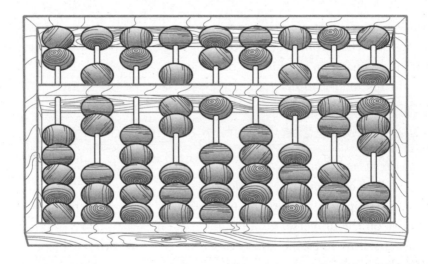

Ancient Chinese mathematics was often about the practical, but it was often combined with the divine too. One illustration of this was Yu's actions to prevent a recurrence of the severe flooding that occurred during the reign of Emperor Shun, his predecessor. (Both men have semi-mythical status in China, partly because we can't say for sure if they even existed.) Yu meticulously studied the flow of the rivers and built a complex system of canals to move floodwater on to fields. Over thirteen years, he is said to have personally overseen the project, sleeping in the same quarters as the farmers and helping in the hard work of dredging the riverbeds.

It worked. Rivers in China's heartlands, among them the Yellow River and the Wei River, flooded no more. This was incredibly important to the many people who made a living by the rivers and used them for travel and the transportation of goods. Life flourished along the riverbanks and Yu attained the epithet 'Great Yu Who Controlled the Waters'. Embarking on the project must have taken great confidence, the sort of confidence you get from spotting a turtle bearing a magic square that promises auspicious times ahead.

The hexagram guide to the galaxy

Mathematics grew alongside political power. The Zhou dynasty, which stretched from roughly 1046 BCE to 256 BCE, was the longest-running dynasty in the history of China. It gave us the philosopher Confucius and the military strategist Sun-Tzu, and the written documents that survive are far more sophisticated than earlier ones. Mathematically speaking, this is also when two of the most influential books of all time first appeared: *The Book of Changes* and *The Nine Chapters on the Mathematical Art*.

The Book of Changes – also known as the *I Ching* – is pretty ambitious in scope. It aims to be nothing less than a comprehensive treatise on the universe, guiding us through how to make the right decisions, how to predict our futures and find our ultimate purpose. Whether it succeeds in doing that is, to say the least, up for debate but culturally and mathematically, its importance cannot be overstated.

The exact origins of *The Book of Changes* are unknown. (Legend has it that the mythical emperor Fuxi, said to be a son of heaven, created it.) The book was probably first compiled sometime between 1000 BCE and 750 BCE and it proposes a form of divination called cleromancy in which yarrow sticks are thrown into the air several times and the pattern they form on landing is interpreted as one of the sixty-four symbols below, now known as hexagrams. Each hexagram corresponds to a chapter in the book, so the pattern formed directs you to a certain text to read and interpret – although interpreting what it meant was far from straightforward.

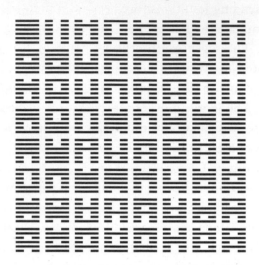

The philosophy contained in *The Book of Changes* is linked to the concept of yin and yang that permeates ancient Chinese culture and says that two complementary halves must come together to produce wholeness. *Yin* comes from the word for the shady side of a hill and *yang* for the sunny side. Yin and yang are said to form the basis of all things, including human beings. It was thought that the dynamics of the world could be understood through this lens with the help of *The Book of Changes*. In the hexagrams, broken lines represent yin and solid ones yang, and the chart displays the sixty-four ways in which the two symbols can be combined in groups of six. For hundreds of years, the great and the good consulted *The Book of Changes* to help them make decisions and understand their purpose in life, and the book took on such importance that all subjects had to embrace it.

In 10 BCE, the astronomer Liu Xin used *The Book of Changes* to interpret observations and calculations to do with the night sky. His Triple Concordance System outlined the movements of the moon, the sun and the planets, and linked them to the sixty-four hexagrams. Although it had its flaws, it was at the time one of the most complex models of the universe yet to be formulated. Liu's system calculated the average length of a lunar month to be 29.5309 days, comparable to the same calculation made by the Maya, and incredibly accurate.

Long-distance sea travel was thriving at this time and the Catholic Church was sending Jesuit missionaries to China to preach, proselytize and disseminate the faith. These missionaries were welcomed by the Chinese. The seventeenth-century emperor Kangxi was especially interested in 'Western Learning' and so invited some of the Jesuits to the court to give lectures on a variety of topics, including mathematics. The Jesuits sent reports back to Rome about the intellectual culture in China, and translations of classic Chinese books made their way back to Europe. However, this two-way traffic displeased the Catholic Church, which had, after all, sent missionaries to teach, not to learn, believing that Christian beliefs were incompatible with Chinese ones. Some missionaries had come to the conclusion that Europe and China had a shared history, ancestry and god, which the Catholic Church saw as blasphemous. So the Catholic Church banned any further learning of Chinese rites.

Despite this, the seventeenth-century German mathematician and polymath Gottfried Wilhelm Leibniz managed to get a hold of a copy of *The Book of Changes*. Reading it, he was astounded to see that the hexagrams were a pictorial representation of a numerical system he had been working on. 'It is a very surprising thing that it perfectly matches my new manner of arithmetic,' he wrote to Joachim Bouvet, the Jesuit missionary who had brought *The Book of Changes* to his attention.[2]

Leibniz's numerical system was born out of the Christian distinction between two states of being: existence and non-existence. He represented these two states as 1 and 0, and came up with a way to represent every number using just these two figures. This approach became known as the binary system. The combinations of 1s and 0s he used for the numbers 0 to 63 were the same as the combinations of solid and broken lines used in the hexagrams in *The Book of Changes*.

Binary, in *The Book of Changes*, was deeply rooted in the philosophy of yin and yang, and Leibniz's binary was deeply rooted in Christianity. Nevertheless, the resulting mathematics was universal and clearly

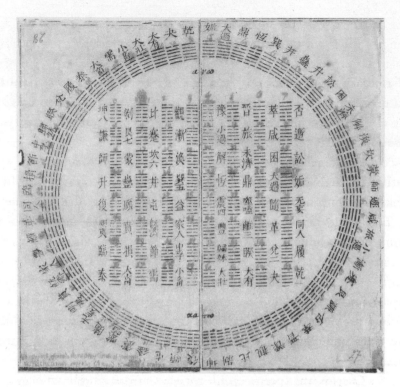

A hexagram from *I Ching*. This was the version that
Gottfried Wilhelm Leibniz saw
in the seventeenth century. He annotated it in ink.

compatible with both Chinese and European culture – and with many others, too. Binary mathematics also appears in the Rhind papyrus, in the mathematics of India in the second century BCE and, at least three hundred years before Leibniz was born, in the counting system of the Mangareva people of French Polynesia.

The origins may have been different, but the binary fundamentals were the same. Mathematics is often intertwined with religion, politics, culture and identity – it is performed by people, after all, so it's hard to imagine it any other way. However, as the case of binary shows, there are many ways to reach a mathematical idea.

Nine chapters that changed the world

The binary interpretation of the chart in *The Book of Changes* didn't feature in the oldest extant versions of the text – it was added later, by the eleventh-century scholar Shao Yong. In times gone by, a mathematician's main job was to preserve knowledge that had already been discovered by making copies of existing works. Many simply copied without engaging with the subject matter, but some couldn't help but make a few improvements and additions along the way, as with binary in *The Book of Changes*. And this also happened with another book from ancient China: *Nine Chapters on the Mathematical Art*.

Nine Chapters is less well known outside East Asia than *The Book of Changes*, but its influence has been seismic. *Nine Chapters* formed the basis for mathematics in East Asia for centuries, rooting mathematics both in problems of a practical nature such as keeping track of time and taxes, and trying to predict what the future might hold through divination.

Nine Chapters was a perfect guide for government officials who had to learn mathematics to manage resources such as grain, labour and time. The first chapter demonstrates how to calculate the area of fields and the second deals with the exchange of commodities. But the level of difficulty goes up quickly. Chapter Eight features problems that contain several unknown quantities, laying the foundations for algebra. The last chapter intricately tackles geometry, featuring problems with 2D shapes such as triangles, rectangles, trapezoids and circles, as well as solid figures such as prisms, cylinders, pyramids and spheres. In the course of the book, the chapters become more abstract and general, although each starts with an illustrative example and then goes on to provide instructions on how to solve more general problems.

The earliest known surviving copy was put together by third-century mathematician and writer Liu Hui. In his opening preface, Liu laments what was lost before his birth and claims that the text of *Nine Chapters* was originally written around 1000 BCE. Historical consensus is that this is too early and that the book was probably first compiled sometime during or after Qin Shi Huang's reign in the third century

BCE. He was the first emperor in the Qin dynasty and arranged for countless books to be burned to avoid comparisons between his rule and that of previous dynasties. Few works survived his wrath.

Regardless of when the book was first compiled, Liu's commentaries on *Nine Chapters* were a feast of mathematics. One particular highlight of Liu's work was his approximation of the ratio of a circle's circumference to its diameter, often written today as pi, or π.* Liu wasn't the first to discover pi, but he found its value to greater precision than before. The Babylonians had known pi was approximately 3. In the third century BCE, the Greek mathematician Archimedes had narrowed it down to a range between 3.140 and 3.142. Liu approximated pi as 3.14159, correct to five decimal places, using the same method (though whether it had somehow made its way to China from Greece is unknown). Today, using supercomputers, we've calculated pi to 50 trillion decimal places. But as astrophysicists need to know the number only to fourteen decimal places to precisely control rockets launching into space, going beyond that doesn't serve much practical purpose. It's more about our love of pi and our keenness to test supercomputers and their algorithms to the limit. For the kind of day-to-day calculations contained in *Nine Chapters*, Liu's number was more than good enough. And the technique he used to arrive at it was a cunning one – involving polygons (many-sided shapes).

The perimeter and the distance from the centre to an edge of a regular polygon are easy to calculate. As the number of sides of such a polygon increases, Liu noticed, the shape comes more and more to resemble a circle. This means that their perimeters and widths become better and better approximations of a circle's diameter and radius. By imagining a regular polygon with 3,072 sides, Liu came up with his approximation of pi.

Following in Liu Hui's footsteps, in the fifth century Chinese mathematician Zu Chongzhi calculated pi even more accurately, by using a 24,576-sided polygon, giving him the correct values to seven decimal places. This remained the world record right up until the early-fifteenth-century Arab mathematician Jamshīd al-Kāshī

* This custom was started by Welsh mathematician William Jones in 1706.

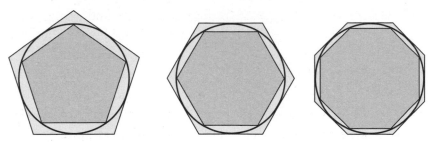

Polygons inside and outside a circle. Ancient mathematicians
would increase the number of sides to approximate pi.

smashed the record by determining pi to sixteen decimal places.
That's as many as we'll ever really need, calculated over seven hundred years ago.

In his version of *Nine Chapters*, in a departure from the usual style
of Chinese mathematics, Liu went beyond the practical problems by
adding mathematical proofs. Instead of looking at illustrative examples in order to demonstrate a broader pattern, he began with the
general, using items from the mathematical-proof toolbox to build
an irrefutable logical argument. These techniques underpin all mathematics today.

One of his proofs was for the theorem often named after Pythagoras but known in China as the Gougu Theorem.* For mathematicians
of this time, triangles were of particular practical use – for example
to calculate the height of an island as viewed from the mainland, the
size of a distant walled city, the depth of a ravine or the width of
a river mouth as observed from a distance. Books such as *Nine
Chapters* commonly featured problems like these as illustrative
examples.

The version of the Gougu Theorem in *Nine Chapters* is the earliest
written statement of this theorem so, arguably, we should be calling
it by that name instead of after Pythagoras. Either way, it is a theorem that has been rediscovered all over the world, including in

---

* *Gougu* is a compound word derived from the Chinese words for two sides of a
triangle. For $a^2 + b^2 = c^2$, a is *gou*, b is *gu* and c is *xian*.

A page from Liu Hui's *Complete Collection of Illustrations and Writings from the Earliest to Current Times*, 1726 edition.

Babylonia, Egypt, India and Greece. Liu's insight was to copy and rearrange right-angled triangles in a similar way to the proof given in the next section. He used a slightly different argument, but many of the underlying principles involving manipulating copies of triangles were the same.

What is a proof, anyway?

The idea of mathematical proof is one that will come up again and again in this book, so it is worth taking a moment to explore exactly what it is.

Proof is how we know that something is true in mathematics. In the strictest terms, to prove a theorem, a mathematician must lay out the starting assumptions – axioms – and the rules of logic that are permissible. Then, using only these axioms and rules, they must fit them together to prove that something is true.

Proofs can often be elegant and beautiful. They can be surprising and joyous. They can also be hacked together, complicated and difficult to follow. There can be more than one way to prove something but, once a theorem has been proved, it is proved for ever.★ As such, the theorems we know from ancient times still stand today.

Take the Gougu Theorem (sorry, Pythagoras!). This states that in a right-angled triangle, where a, b and c are the lengths of its sides, with c the longest side, $a^2 + b^2 = c^2$.

There are more than a hundred known ways to prove this today. Here is a relatively modern one:

Take four examples of this triangle

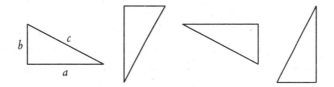

then rearrange them into a square with a hole in the middle.

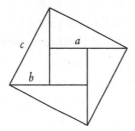

There are two clear ways to calculate the area of the large outer square (formed of the four triangles and the square in the middle). The first is to simply multiply the length of one side by itself to get c^2.

★ Of course, mistakes are occasionally found in proofs, but this raises the question of whether it really was a proof in the first place.

The second method is to add the area of the smaller square in the middle to the areas of the four triangles. The smaller square has sides of length $(a-b)$ so its area is $(a-b)^2$. The area of a triangle is half its height multiplied by its width. So each triangle has an area of $\frac{1}{2}ab$. Since there are four of them, in total the area of the triangles is $2ab$.

Putting this together, this means we have two ways to express the area of the large square, so

$$c^2 = (a-b)^2 + 2ab$$

If you then multiply out the brackets, this gives

$$c^2 = a^2 + b^2 - 2ab + 2ab$$
$$= a^2 + b^2$$

Et voilà! Or, as some mathematicians like to write when proving a theorem, QED (from the Latin phrase *quod erat demonstrandum* meaning 'what was to be shown').

That the Gougu Theorem has stood strong all these years is both extraordinary and completely normal in mathematics. If you compare this to ideas in any other scientific discipline, it shows just how powerful a tool it is. Almost every belief from other sciences has been rewritten in one way or another over time. This is by design, as the scientific method requires iterative improvements on what has gone before. Scientific theories do not stand for ever; they stand only until a better one comes along. In mathematics, once something is true, it is true for good.

However, this also shows the difference between truth in the real world and mathematical truth. Mathematics underpins many scientific theories. These theorems have been proved, yet the science based on them sometimes turns out to be wrong. That is because the intersection between mathematics and the real world is a complicated one. We don't know what the starting assumptions of the universe are, nor what logic is permissible. Within the mathematical universe we create, the theorems we prove are true for ever, but it is not always obvious to what extent the mathematical universe corresponds to the universe we live in.

None of this is to diminish the power of mathematics to describe the real world. Nothing – from quantum physics to the study of cells – would be the same without mathematics and its proofs. But as many ancient mathematical cultures show, proofs are not a necessary condition of mathematical prowess. Chinese mathematics only occasionally featured mathematical proofs, and yet ancient China was a mathematical powerhouse. The rulers valued mathematics because it was linked to their authority to rule. And so the person responsible for teaching mathematics to those in the emperor's closest circle was a very important person indeed.

Lessons for women

In 202 BCE, sometime after *Nine Chapters* was written, the Han dynasty came to power, and under the Han there was great technological advancement. Paper and improved versions of sundials and water clocks came in this era, as did Liu Xin's calendar system. So, too, did an extraordinary historian and mathematician by the name of Ban Zhao, one of the first known female mathematicians in the world.

Ban Zhao was born around the year 45 BCE into a well-known family of scholars. Her father, Ban Biao, was a historian and the original author of what was to become the *Book of Han*, the official history of the Han dynasty from 206 BCE to 23 CE. However, her father died before he completed the project and when she was still a young girl. Her older brother, Ban Gu, took over the task. At the age of fourteen, Ban Zhao married a resident of Anling (near modern Xianyang) called Cao Shishu, but her husband did not live long after their marriage. Widowed, she would dedicate the rest of her life to scholarly work. She had been taught how to read and write by family members and given an education in the teachings of Confucius. His virtues and philosophies were cherished by the ruling class and were the basis of the notorious civil-service exam, designed to mark out the intellectually gifted. Her knowledge put her in good stead with Emperor He of Han's court. He recognized her intellect and gave her the job of teaching mathematics and astronomy to Empress Deng Sui, He's second wife, and the imperial concubines. Ban and Deng may have been the first teacher–pupil pair in the study of mathematics in China.

We do not know exactly what was taught, but we do have some evidence to go on. Ban was well versed in the Confucian classics and would have read *The Book of Changes*. The imperial family's library, the Dongguan Imperial Library, housed books on a variety of subjects, including mathematics and astronomy. As she was born after Qin Shi Huang's reign, during which many books were burned, the written documents that she was able to read would all have been relatively new. Her brother had initially made good progress on the *Book of Han* and had broadened the scope of the book to include the first two hundred years of the Han dynasty. However, he was so meticulous about recording every detail, including mistakes made by officials and emperors, that the court started to become suspicious of the work he was doing. He was accused of altering history and sent to prison, and later accused of plotting a coup against the emperor. Soon afterwards, he died in prison.

Ban Zhao, then in her mid-forties, was given the hazardous task of taking over from her brother. She spent decades working on the *Book of Han*, paying particular attention to Emperor He's mother and including the familial histories of women associated with the court, at a time when women were rarely included in such histories. She also probably added a chapter on astronomy featuring interpretations of the movement of stars, eclipses and the weather during the Han dynasty. Ban's teaching and position in the court won her considerable political influence. Empress Deng became regent after the death of her husband and often asked Ban for advice on important matters of state. At the imperial court, the men called her the Venerable Madam Cao in recognition of her talent and achievements – a name derived directly from her husband's. The women of the court preferred to call her 'the gifted one'.

In the course of her life, Ban noticed that there was little written specifically for women in Confucius's teachings in society and in her

惠班名昭一名姬博學高才遠曹世叔見固與續書
曹大家班惠班君熊
未及竟而卒和帝詔昭就東觀藏書閣踵而成之數召入宮令皇后
諸貴人師事焉號曰大家

A later rendering of Ban Zhao by Jin Guliang (1690).

sixties she set out to correct that. Her hugely influential *Lessons for Women* outlined seven simple rules for women to follow to maintain proper conduct in Chinese society.

It began, 'I, the unworthy writer, am unsophisticated, unenlightened, and by nature unintelligent.' Of course, she was none of these things, but Confucianism put a high value on modesty. Many of her readers would have seen such a declaration as giving the impression of the opposite, almost like a humblebrag today.

On first glance, many of the seven rules are very much of their time, revolving around being a subservient wife and obeying one's husband – traditional pillars of Confucianism. Yet, if taken in the context of her life and times, it can be read as a guide for women on how to get by in a patriarchal society. Among the rules is a strong advocacy for literacy and education for women – the first Chinese text to espouse such a view. Over a thousand years later, during the

Ming and Qing dynasties, learned women would turn to Ban Zhao's words on education to bolster their arguments for gender equality.

A band of rebel geometers

Over the course of many hundreds of years, Chinese mathematics became a force to be reckoned with. When you look back at the body of work, it is as formidable as any produced anywhere else on the planet at any other time in human history. Prosperous emperors used their resources to support mathematicians, especially to help uncover favourable truths about their rule. Chinese mathematics had a particularly strong influence in countries in East and Southeast Asia such as Japan, Korea and Vietnam. These former tributary states adopted the Chinese writing system and followed the practical, example-led style of writing mathematical textbooks. This example-led method of mathematics was integral to Chinese mathematics but, for a brief period in the fifth century BCE, it looked as if a different approach could take over.

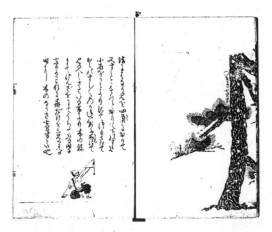

Two pages from the 1627 Japanese mathematics text *Treatise for the Ages*.

Mozi was a philosopher who lived during the Hundred Schools of Thought period in China, where new ideas came along at a rapid pace. In the mid-4th BCE he founded a school that took aim at the

Confucian ideas he had been taught, advocating a more meritocratic society rather than one based on social class. As part of this rethinking, he and his followers, the Mohists, wrote on geometry in a way that was completely different to what had gone before. Rather than starting with specific examples, they began with general assumptions and then combined them into one logical system in order to prove the properties of points, lines and shapes. Specific examples were then plugged into the general theory. This went further than Liu had gone when proving the Gougou Theorem – not only was Mozi using arguments, he was also looking at the assumptions that underpinned them.

Mohism was, for this brief period, very popular, and it could have led Chinese mathematics in a different direction. It died out in Qin China as Confucianism became dominant again, but a similar approach would independently rise to prominence elsewhere. A different mathematical tradition was developing thousands of miles away that would put building proofs from general principles at its centre.

3.   A Town Called Alex

The year was 415 CE. Alexandria, a city in the Eastern Roman empire in what is now Egypt, had the largest population of any city in the world, with somewhere between 300,000 and half a million inhabitants. Its port, Alexandria, connected it to Europe and the Middle East. The city was a bustling meeting point for the brilliant minds of the day. It was the de facto intellectual capital of the Nile Delta, the Mediterranean and the Western Desert of Egypt.

And the city was on something of an intellectual roll. The ruling elite greatly supported libraries and museums, enabling Alexandria to produce generations of outstanding philosophers, astronomers and mathematicians. This support wasn't in the service of some noble ideal of promoting knowledge for the sake of knowledge – rulers wanted to bolster their legacies, and building vast collections of knowledge was a clear sign of power. Power was never far from view in Alexandria.

Euclid of unknown

Alexandria was founded when Alexander the Great conquered Egypt in 332 BCE. He arrived there with an army of Greeks and Macedonians, but they were barely required to fight as the Egyptians were already eager to kick out the ruling Persians. Alexander set about building the Greek-style city that would bear his name. After his death, one of his generals, Ptolemy I, declared himself pharaoh of Alexandria around 300 BCE and fused major gods from each culture to form a universal one, Zeus-Ammon, to signal the fusion of Greece with Egypt. He also declared Alexandria the capital of Egypt. Alexandria forged links with cities such as Athens, sharing the most up-to-date knowledge, including about mathematics, as scholars travelled between them.

Mathematics, for the ancient Greeks, just as for other cultures, was about more than practical calculations. Many Greek mathematicians believed that mathematics encapsulated a divine form of beauty and was a gateway to understanding the reason for human existence. As Plato wrote in his *Republic*, 'geometry will draw the soul towards truth.' Mathematics was about uncovering 'knowledge of the eternal'.[1]

Some mathematical ideas were revered over others because of their supposedly magical properties. The Pythagoreans – a mathematical cult comprising apparent followers of Pythagoras – believed 10 was the most perfect number. They also believed 1 was the origin of all things because it could be used to generate every whole number through repeated additions $(10 = 1+1+1+1+1+1+1+1+1+1)$. Many of these beliefs gave extra weight to the pursuit of mathematical discovery, but they also hampered the ability to follow the logic of mathematics through to its conclusion. The mathematician Iamblichus, for example, was an early thinker on the number 0, but the number never made it into mainstream Greek mathematics because it didn't fit with their world view. As Aristotle once wrote, 'There is no void existing separately, as some maintain.'[2]

Aristotle, Plato and Pythagoras were among those who contributed in the early days of the Greek mathematical scene, but it was Euclid whose work has arguably had the biggest influence. Very little is known about Euclid's life. He was probably born around 325 BCE, but so few surviving contemporary sources mention him we aren't even sure where. He is often referred to as Euclid of Alexandria, which certainly implies he lived there, at least. There are scholars both before and after him from the same region who had extensive biographies written about them, but in his case almost everything we think we know was written long after his death.

Our best guess is that Euclid came to Alexandria after Ptolemy I declared himself pharaoh. Ptolemy funded and supported the gathering of knowledge. He founded the Musaeum, an institution that became the centre of learning for the region and was home to the famous Library of Alexandria. The library was one of the largest in the world, intended to show off the vast wealth of Egypt. It housed hundreds of thousands of scrolls and had many reading, dining and meeting rooms, as well as gardens and lecture halls. Euclid was one of the first scholars affiliated with the library and the Musaeum. And that's about the extent of our knowledge of Euclid, the person.

The Library of Alexandria, a nineteenth-century artistic rendering based on archaeological evidence.

The lack of information about Euclid's life is in stark contrast to our knowledge of his work, which has not only survived but has become foundational to modern mathematics. His *Elements*, a thirteen-book treatise on mathematics, is one of the most influential works ever produced. Its influence on Europe and North America has been similar to that of *Nine Chapters* in East Asia.

Most of the *Elements* is taken up with geometry, covering both two-dimensional and three-dimensional shapes, but it also features some number theory – the study of numbers and their properties. What was so innovative is its strict adherence to the basic principles of proof. Many of the results it includes were known before Euclid's time, but he managed to put them into one coherent framework, clearly laying out his assumptions – known as axioms or postulates – and using them to build logical arguments.

One of the oldest surviving fragments of Euclid's *Elements*,
written on papyrus. Dated *c.*100 CE.

Euclid begins the first book of the *Elements* by defining a point as 'that which has no part'. He then goes on to list four further axioms, among them 'a line is breadthless length'. From these fundamental geometric definitions, the rest of his work follows. Euclid's method – starting with the simplest axioms and building up to more sophisticated ideas and theorems – has become a central tenet of mathematics, so much so that the *Elements* formed the basis of teaching mathematics in Europe right up until around the middle of the twentieth century.

This approach is different to the one taken in *Nine Chapters*, which is organized far more around practical examples as a means to demonstrate general principles. Neither one of these perspectives should be considered inferior to the other. The axiomatic approach of the Greeks has often been lauded, but rooting understanding in real-world applications is an equally valid approach.

The *Elements* offers a masterclass on how to construct a mathematical argument but, as thousands of years of students can vouch, that is not always an easy task. Though perhaps apocryphal, a conversation between Euclid and Ptolemy I gives a taste of later students' woes. Struggling with the difficulties and sheer length of the *Elements*, Ptolemy asked if there was a shortcut to master mathematics, to which Euclid replied, 'There is no royal road to geometry.'[3]

Alexandria, the great

Beyond Euclid, Alexandria was home to its fair share of important mathematicians, and one with a particularly interesting life story is Pandrosion. As we will often see in this book, assumptions made by historians are often perpetuated by those that follow, giving a false impression of someone or their life. This is very clearly the case in Pandrosion's story.

Much of Pandrosion's life – which began around the year 300 CE – was either never recorded or has long since been lost, but we can be fairly sure that she came up with an approximate method for one of the most confounding problems of ancient geometry: doubling the cube. Given the length of the edge of a cube, how do you construct another cube with double the volume?

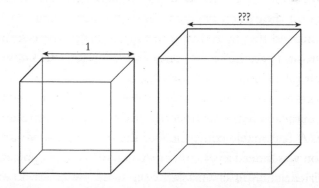

Today, we can use algebra to easily find the length of an edge in the doubled cube,* but ancient mathematicians didn't have this luxury

---

* Answer: $\sqrt[3]{2}$, although the ancient Greeks didn't know that.

available to them. They were trying to find a way to construct the second cube using just a straightedge and a compass. This sort of technique is common throughout geometry – the idea being that with these two tools you can create circles and lines based on the shapes you start with. Though you can use physical tools to re-create the process, Euclid included postulates about these tools in the *Elements*, which meant that mathematicians could then imagine theoretical versions of the tools and what might be possible with them.

Or, in this case, not. Perfectly doubling the cube using only a straightedge and a compass is in fact impossible (although this would not be known for certain until the nineteenth century, when Pierre Wantzel proved it using modern algebra), so it's no wonder the problem hung around for so long.

Pappus was a contemporary of Pandrosion, and a resident of Alexandria. In the *Collection*, his eight-book epic on mathematics, he alludes to Pandrosion's technique for approximating the solution – but only to make light of it during a sarcastic rant. His main criticism of her work is pedantic in nature. According to Pappus, Pandrosion and her students had confused a 'problem' with a 'theorem'.[4] He wrote that a problem is something that can either be true or false, whereas a theorem must be proved to be correct. It seems very unlikely that Pandrosion and her students hadn't known this distinction, and so this may be one of the earliest recorded cases of mansplaining. Although it is often assumed that there was some sort of rivalry between Pappus and Pandrosion because of his comments, in reality we know so little about their relationship it is impossible to tell.

We do know, however, that Pappus used the female form of address and grammatically feminine adjectives for Pandrosion throughout the *Collection*. A reasonable conclusion to draw from this would be that Pandrosion was indeed a woman, but when historian of ancient mathematics Friedrich Hultsch translated the work from Greek into Latin in the nineteenth century he took the opposite view, assuming that presenting Pandrosian as female must have been a mistake. Many historians then followed Hultsch's lead. It wasn't until 1988, when Alexander Jones produced an English translation of the *Collection* and put forward the case that Pandrosion was a woman, that the prevailing view started to be challenged. Other scholars have since substantiated the idea, and it is now commonly accepted. This makes Pandrosion the earliest known female mathematician in this part of the world – an accolade previously reserved for the better-known mathematician Hypatia.

Hypatia the teacher

Female mathematicians in ancient Greece were not unheard of, although many have been lost from the historical record. In the seventeenth century, scholar Gilles Ménage was able to collect references to sixty-five women from around this period. To name just a few, there was Aspasia (*c.*470–*c.*400 BCE), who cultivated one of the most prominent intellectual salons of the time, drawing to it renowned philosophers such as Socrates. Her name appears in Plato's writings, and he is said to have been impressed by her intelligence and wit. Arete of Cyrene (*c.*400–*c.*300 BCE) wrote over four hundred books on philosophy and natural science, although, unfortunately, none has survived. And then there was Hipparchia (b. *c.*325 BCE), who wore men's clothes, lived in equality with her husband and wrote several works on philosophy – although, again, they have been lost to time.

These examples should not give the impression that ancient Greek society was an egalitarian utopia. It was still male-dominated. In the case of Hipparchia, people were probably shocked by her and her husband's lifestyle. Men and women rarely worked together in public. And we can't be sure just how much our knowledge of these women has been distorted by those telling the story. We've kept to the most basic of facts in the previous paragraph, but as Kathleen Wider wrote in the 1980s in a paper about female philosophers in the ancient world, both ancient and modern sources are 'sexist and easily distort our view of these women and their accomplishments'.[5]

As we will see, history has also been unfair to Hypatia. She was born around 350–370 CE. Nothing is known about her mother, as she does not feature in any surviving sources, but her father was the prominent mathematician Theon, the head of a school called the Mouseion, which specialized in astronomy and mathematics (and was separate from the Musaeum of Alexandria). He was perhaps best known for updating the *Elements*.[6] His revised edition, with commentary, seven hundred years after the original, became the most widely read version of the work for the next few centuries.

It is thought that Theon once said to Hypatia, 'Reserve your right to think. For even to think wrongly is better than not to think at all.' She soon excelled in mathematics. As Edward Watts, one of her biographers, puts it, she 'quickly proved more capable than her father and developed competencies greater than his', and so moved 'from being a mathematics student at her father's school to being one of his colleagues'.[7]

We don't know exactly why Hypatia would have been able to go down this unconventional route. It was common for daughters in learned families to be given an education as part of their preparation to be married off into another learned family, say to a male philosopher or rhetorician. Some women, particularly in the upper classes, might continue their education, pursuing a formal study of philosophy, which then would have included astronomy, geometry, arithmetic and the texts of Aristotle and Plato. However, becoming a mathematician and a teacher at a school, as Hypatia did, wasn't normally part of the programme.

Hypatia's earliest works were commentaries on mathematical texts. These texts contained little in the way of explanations, as so much was discussed and passed down orally. This naturally bred confusion, so mathematicians often needed to use their expertise to decode what was written and present it in a way that was easier to understand. When a skilful mathematician made significant improvements to a text, it prolonged the life of the original. Back then, commentaries were the way in which scholars proved themselves to be first-rate mathematicians. Hypatia produced commentaries on many mathematical and astronomical texts, including Diophantus's *Arithmetica*, Apollonius's *Conics* and Ptolemy's *Almagest*. These were foundational texts of what would later become known as algebra, geometry and astronomy respectively.

Arithmetica was the first book in history to use symbols for unknown quantities,* in the way we often use x and y today. For example, Diophantus wrote

$$\mathrm{K}^{\upsilon}\overline{\alpha}\ \zeta\overline{\iota}\ \pitchfork\ \Delta^{\upsilon}\overline{\beta}\ \mathrm{M}\overline{\alpha}\ \acute{\iota}\sigma\ \mathrm{M}\overline{\varepsilon}$$

which in today's notation we might write as

$$x^3 - 2x^2 + 10x - 1 = 5$$

Diophantus's notation didn't go as far as being fully algebraic. The symbols were shorthand, convenient for writing out mathematics concisely, but mathematicians of this time didn't see symbols as things that could be mathematically manipulated. That leap would come from a ninth-century mathematician in Baghdad who we will meet in Chapter 6. However, Diophantus's notation was a step in the direction of making it possible to leave gaps in calculations to be filled in later and to express extremely general mathematical relationships. Without a way to do that, there could be no $E = mc^2$.

Hypatia's addition to *Arithmetica* was to include several student exercises. She must have seen the mathematics in the book and felt that her students needed a bit more practice on simultaneous equations – equations where there are multiple unknown quantities. As such, she added the following equations to be solved:

$$x - y = a$$
$$x^2 - y^2 = \left(x - y\right) + b$$

where a and b are known.[8] Where this equation came from is unclear, but the point of it is evident: solving simultaneous equations would have been a particularly useful skill for students in her school studying astronomy, where they often came up.

Hypatia also revised commentaries made by her father on Ptolemy's *Almagest*. We know this because Theon himself acknowledged Hypatia's work. In it, he wrote 'the edition having been prepared by

---

* Though they may have only been added in later editions.

the philosopher, my daughter Hypatia'.[9] The book was particularly useful for teachers and, as Hypatia was a teacher herself, it seems probable she wrote it partly for her own benefit. It assumed no prior knowledge of astronomy or the *Elements*, and explained, to the best of the ancient Greeks' knowledge, how the solar system worked.

At the time, people still believed that the sun revolved around the Earth, and Ptolemy had put forward a calculation that would describe how far the sun moved in relation to the Earth each day. This required a form of long division, and Hypatia improved upon it, proposing a new method involving tables to simplify the process.[10]

Hypatia edited a version of Apollonius's *Conics*, which explored shapes such as circles, ellipses, hyperbolas and parabolas, each of which arises from intersecting a plane with a cone.

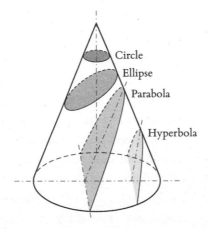

Her work on the book helped Apollonius's ideas survive until they were re-examined in the seventeenth century, with such books appearing to partially mimic her style. Astronomer Johannes Kepler was one of those who studied Apollonius's *Conics*, and this may have been where he learned of ellipses – the shapes he would use to describe the orbits of the planets in our solar system.

Hypatia's interests extended to scientific instruments, and she helped to fine-tune the design of the astrolabe, a device used to predict the future position of the stars and the planets. Astrolabes had existed long before Hypatia, but she had learned a lot about them from him and his book *Treatise on the Astrolabe*, possibly the first book to mathematically describe the instrument. When one of her students, Synesius of Cyrene, wanted to create a plane astrolabe, a smaller and more compact version of the original, Hypatia taught him how to do it.[11]

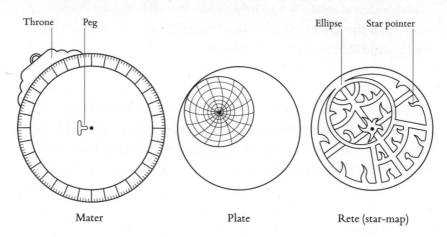

The mechanism of a plane astrolabe. Hypatia helped her student to craft this.

On another occasion, Synesius asked Hypatia to help him build a hydrometer, an instrument used to measure the density of liquids. It consists of a float and a container for the liquid with a scale on one side. The letter between Synesius and Hypatia is the oldest known reference we have to this instrument. Though some have read the letter as suggesting that Hypatia invented the device, others say it is more likely that she simply had the expertise to replicate it.

By the time Hypatia turned thirty, she had become a prominent intellectual figure in Alexandria. Her father was now around the age of fifty-five and decided to take a phased retirement, so she took over his school. Presumably, he felt secure in this decision, knowing that, with his daughter, the school would be in excellent hands.

Though there had been other learned women in Alexandria, Hypatia was the first we know of to consolidate her social status as a renowned female intellectual. People travelled to Alexandria to hear her speak. Her home became a gathering place for educated people to discuss and study a blend of philosophy, mathematics and astronomy. She herself was pagan, but she welcomed people of all religions.

However, life was not easy for learned women. Female teachers had to endure gossip about their sex lives as they often taught young, unmarried men in private. Hypatia's virginity was a topic of discussion and suspicion among the people of Alexandria. Although it's possible that she married a philosopher named Isidore, it seems more likely that she lived a life of celibacy and rebuffed approaches made to her by students, apparently calming them down with music.

By the 380s CE, the majority of people in Alexandria were Christian, and in the following decades the city became divided between those who were Christian and those who were not. The city became more polarized and religious views more extreme. In 412 CE, the bishop of Alexandria, Theophilus, died, and a bloody power struggle ensued. His nephew Cyril won control of the city and began to clamp down on all who had opposed him. He took particular umbrage with Jewish people, expelling many from the city and closing down synagogues. The Roman prefect Orestes was furious at Cyril's behaviour and sent a scathing letter to the imperial court in Constantinople outlining the situation. Cyril initially wanted to make peace with Orestes, but Orestes refused, putting him at risk of Cyril's wrath – which in turn put Hypatia at risk too.

Hypatia was sceptical of Cyril, but it is not clear that she took any particular side. Public intellectuals were generally considered above political squabbles, so there is little reason to think she would have been involved. Intellectuals were often sought out for advice, however, and she regularly provided counsel to Orestes. Cyril began a campaign to ruin Hypatia's good name and standing. Rumours abounded that she had been the one to prevent Cyril and Orestes from becoming allies. Later, the seventh-century bishop John of Nikiû would write that 'she had beguiled [Orestes] through her magic'.[12] Declaring that a woman was some sort of witch was a terrible yet effective technique for turning public opinion against her, as it has been throughout many periods in history.

Cyril summoned an unlikely cohort to threaten Orestes: five hundred Christian monks. In Alexandria, monks vandalized property and on rare occasions even tortured and executed people. Setting a group of monks on an opponent was a common intimidation tactic. At one point, one of Cyril's mob of monks, amid the frenzy, struck

Orestes on the head with a stone. Orestes survived the blow, and had the monk tortured and killed.

This was enough to push the city into outright chaos. An angry Christian mob formed in the city streets, pawns in the political feud playing out in the city. The mob was bent on destruction and happened upon Hypatia travelling in a carriage. They dragged her into a nearby building and stabbed her to death with broken pieces of pottery, then paraded her body through the streets and set it on fire. This horrific attack was in line with an Alexandrian punishment reserved for the very worst criminals. But Hypatia wasn't a criminal. She had contributed so much to Greek mathematical knowledge. Through her books, commentaries and teaching she had demonstrated that she was a mathematician of the highest calibre. The power struggle between Orestes and Cyril had spilled on to the streets, and Hypatia, caught in the political crossfire, paid with her life.

Looking back

After Hypatia's murder, many scholars associated with her school left Alexandria and moved to Athens. Some had only been living in Alexandria because of Hypatia in the first place; others wanted to escape the consequences of Cyril's rise to power. In Athens, pagan intellectuals retold her life story, recounting how she had been murdered and putting it in the context of pagan persecution in the name of Christianity. Hypatia was treated as a pagan martyr. In the fifth century, church historian Socrates Scholasticus wrote her story, blaming and shaming the Christian people for what they had done. In contrast, some Christian writers defended Cyril and the killing. John of Nikiû labelled the mob 'a multitude of believers in God'.[13] It was described as an act of 'ritual civil cleansing',[14] and that killing Hypatia was justified as part of the effort to expunge paganism from the city.

Of course, it was not just Hypatia's role in religious disputes that interested subsequent historians. Influential scholar Damascius retold her story in *Life of Isidorus* in around the fifth or sixth century CE, poking fun at the traditional philosopher's *tribon* cloak she wore. He made no such quips about the men in his work.[15] Seventeenth- and eighteenth-century writers also revisited the life story of Hypatia. John Toland's book *Hypatia* from 1720, for example, introduced Hypatia as a model woman, and he encouraged contemporary women to become educated professionals like her. Hypatia's life was honoured and praised, but Toland used the story to argue against Christianity. Opposition to Toland's work soon followed. The Christian writer Thomas Lewis called Hypatia 'a most impudent school-mistress of Alexandria',* criticizing Toland's work and defending Cyril's actions.

In the Victorian era Hypatia again proved a compelling character for writers, this time for Anglican priest and Christian socialist Charles Kingsley. He wrote a novel called *Hypatia* that was translated into several European languages and widely read. However, Hypatia's mathematical talents were quietly excised from the story. Instead, the novel focused on her body. At the climax of the novel Hypatia converts to Christianity and is stripped naked and murdered by monks under the image of Christ. The novel was adapted into a play, *The Black Agate, or, Old Foes with New Faces*, in 1859 and the story spilled over into other visual arts, inspiring a flurry of nude paintings of Hypatia.

Later nineteenth-century works depicted Hypatia wearing her philosopher's robe. This shift away from sexualizing her continued into the twentieth century, with historians starting to unpick the damaging fictions that had been told and retold about her. Maria Dzielska's 1995 biography *Hypatia of Alexandria* reconstructed her life story from documents that existed in Hypatia's time. As Dzielska wrote, 'Hypatia has become a symbol both of sexual freedom and of the decline of paganism – and, with it, the waning of free thought, natural reason, freedom of inquiry . . . For those who choose to restrict their focus to the actual historical sources, it is possible to sketch out a clear profile of Hypatia, undistorted by ahistorical idealization.'[16] Dzielska checked the facts and pieced together

---

* The full title of Lewis's book was the outrageously long *The History of Hypatia, a Most Impudent Schoolmistress of Alexandria. Murder'd and Torn to Pieces by the Populace, in Defence of St Cyril and the Alexandrian Clergy. From the Aspersions of Mr Toland* (1721).

Hypatia by Julius Kronberg, 1889 (*left*), and by Alfred Seifert, 1901 (*right*).

Hypatia's life as a philosopher and mathematician, removing the various ideological biases that had been etched over one another for centuries.

In 1998, Hypatia was included in the encyclopaedic book *Notable Mathematicians from Ancient Times to the Present*. Instead of a nude image, it features a portrait, and her entry begins with 'the earliest known woman mathematician, wrote commentaries on several classic works of mathematics'.[17] Although she is no longer the earliest known woman mathematician – that honour goes to Ban Zhao – this simple entry is reflective of the scant details we have of Hypatia beyond her brutal demise. But as Edward Watts wrote in his biography: 'Hypatia's heroism lies not in the brutality that she suffered at the end of her life, but in the subtle barriers she overcame each day while she lived.'[18]

4.   The Dawn of Time

So far, we have been working our way through the history of mathematics by putting our focus on one location and period of time before moving on to the next. This works well for understanding individual slices of ancient history, but for some parts of mathematics we need a broader approach. Such is the case with the mathematics of time. So let us, for a moment, skip forward a little.

It was April 1883 and there was an extraordinary meeting of the senate at the University of Mumbai,* India. Around forty professors, city officials and judges attended to discuss an important matter, one that would determine the very fundamentals of public life in India for years to come. It all centred around an apparently simple mathematical question, but one that nonetheless was so controversial it caused protests and riots.

That question: what time is it?

A few years earlier, construction had been completed of an 85-metre-tall clock tower at the university in the style of that which houses Big Ben in London. The Rajabai Clock Tower was a striking addition to the skyline, but it was a clear symbol of British colonialism and, as such, the time it would display was not just about hours and minutes but about politics and power too.

For most people in Mumbai, as in many other places in the world, timekeeping was tied to the sun. Clocks on 'Bombay time' were based on the rising and setting of the sun, which was perfect for people going about their daily business. But with the rise of trains and telegrams, geography wasn't what it used to be. People and information could now be transported from one place to another much more quickly, and cracks in the old way of doing things were becoming apparent. Different cities

---

* Then known in English as Bombay University.

in India, such as Chennai and Kolkata,* were in different time zones, with some only minutes apart, depending on their geographical positions.

After previous failed attempts, a universal Indian Standard Time was proposed towards the end of the nineteenth century and was set to be imposed in 1906. However, many people resisted the proposal. Some resented the fact that the edict had been imposed by colonial Britain; others didn't want their workday to be ruled by the time the sun set in another location. The concept of time, and how to measure it, was something that affected people's everyday lives – so much so that a few years after the university senate meeting, thousands of cotton-mill workers rioted as a result of this temporal conflict. The Battle of the Clocks lasted for decades.

Clock time had not always been so important. The idea that some-one should track their day in increments as small as seconds, minutes and hours is a relatively recent phenomenon, as is the idea of syn-chronizing these measurements across geographical boundaries. But for as long as humanity has existed, people have noticed the rise and fall of the sun, the waxing and waning of the moon, and the chan-ging of the seasons. These observations led to the production of the first calendars and a more sophisticated understanding of the cosmos. And it all depended upon mathematics.

Cosmic calendars

Many of the mathematical civilizations discussed in this book so far developed their own calendars, primarily to track the stars in order to help people plan and make decisions. The Babylonians, much like the ancient Chinese, believed that everything that happened on Earth could be explained by what happened in the sky and hired court astrologers to give them advice when they had important decisions to make. These astrologers were part diviner, part mathematician. A clay tablet dating from somewhere between 350 BCE and 50 BCE shows Babylonian astrologers calculating how far Jupiter travelled in

* Then known in English as Madras and Calcutta.

a period of time using a technique that was previously thought to have been invented in Europe in the fourteenth century. By estimating the area under a curve using four-sided shapes called trapezoids, they could better follow Jupiter's motions, which they then used for predicting the weather, plagues and the price of grains.

Babylonian astronomers meticulously recorded the movement of many celestial bodies on clay tablets, using astrolabes, among other tools, for precision. Their calendar, much of which was inherited from the Sumerians, along with their number system, was based on twelve lunar months, each beginning when a new crescent moon was first sighted low on the western horizon at sunset. Astronomers periodically added leap months to the calendar when discrepancies between the calendar and seasons became obvious. This meant that the calendar year remained close to the solar year, making it more accurate than previous calendars.

The calendar was refined early in the third century BCE by Kidinnu, a Babylonian astronomer and mathematician. Astronomers before him had believed that the speed of the moon was constant, but Kidinnu improved on earlier methods for calculating the position of the moon and the length of time between two full moons and found that its speed changed periodically.* With this knowledge, the Babylonian calendar became more reliable, and it was soon picked up on by astronomers in Alexandria and Rome.

The first signs of a calendar system in China arose during the Shang dynasty (1600–1046 BCE). The Chinese calendar followed the lunar cycle and consisted of twelve months, each with thirty days, making a 360-day year. The calendar took into account both lunar and solar cycles and was adjusted for the seasons. Emperors would then issue decrees based on the calendar, such as when to attack during war or when rituals should be conducted.

Over in Mesoamerica, starting from around 800–500 BCE, the Maya lived by no fewer than three calendars. The first, the Tzolk'in, was used primarily for rituals and in scheduling religious events but also to diagnose illnesses and to make important decisions relating to business and

---

* This is because its orbit is an ellipse rather than a circle.

harvesting. It was a 260-day calendar consisting of thirteen months of twenty days. We don't really know why these numbers were chosen, but one suggestion is that the Maya had thirteen gods and twenty was a significant number that represented humans (ten fingers, ten toes). Another is that in the region there are 260 days and 105 days between the two days when the sun is directly overhead.

The second calendar, the Haab', consisted of 365 days and comprised eighteen months of twenty days each, plus a single five-day mini-month. The months of the Haab' were named after agricultural and religious events. If a person told you their birthday, or their wedding anniversary, they used the Haab'. The Maya would often combine these two calendars, giving dates both in the Tzolk'in and the Haab'.

The third way in which the Maya measured the flow of time was with the 'Long Count'. This was a 5,125-year cycle measured from what they believed to be the dawn of time (12 August 3113 BCE) and features on many Maya monuments, alongside symbols to represent the Tzolk'in and the Haab'. You may remember from the doomsday-filled early 2010s that some people believed the ending of a Long Count cycle on 21 December 2012 meant the Maya had predicted the world would end on that date. The Maya in fact believed that this was just one of many cycles of rebirth and that things would continue as they were. Unsurprisingly, that is exactly what happened.

A deity a day

The Babylonian, Chinese and Maya calendars are certainly some of the oldest we know of. The Egyptians, too, were recording the passage of time early on with a 365-day year split into three seasons of 120 days each, along with five additional days. However, in 2019, with the discovery of the magnificent limestone sanctuary, Yazılıkaya, in what is now Turkey, the story of another potential ancient calendar came to light.

Yazılıkaya's tall rock walls point boldly to the sky, as was the intention of those in the Bronze Age Hittite empire that built it sometime before the late 1500s BCE. The walls are adorned with sculptures, depictions of deities and symbols, and together form an open-air retreat. The structures have stood for more than three millennia, but until recently their purpose was unknown.

The Hittites had a well-organized military that regularly fought with the Kaskas, a tribe who lived to the north along the coast of the Black Sea, as well as with the New Kingdom of Egypt and the Assyrian empire to the south. However, eventually, the Hittites learned how to make peace with many of the opposing groups in the region, becoming known more as diplomats than fighters. They wrote in widely spoken Akkadian when communicating with others to indicate that they didn't mean to make any trouble.

One particularly renowned diplomat was the thirteenth-century BCE Hittite queen and priestess Puduḫepa, known for the peace treaties she wrote and marked with her seal. She reigned with her husband and was closely involved in political and religious affairs. She decided to organize the Hittites' many gods into a pantheon of the deities she deemed to be most important. Puduḫepa made the sun goddess, Arinna, the queen of gods. This hierarchy was on display at Yazılıkaya and so it is now thought that the site was considered one of the holiest places in the Hittite kingdom. Recently, archaeologists have also begun to think that Puduḫepa's pantheon could have been used to mark the passing of time.

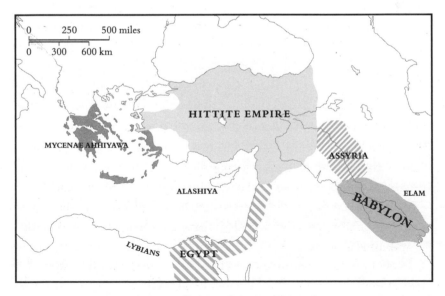

Hittite civilization (*c.* 1600–1180 BCE).

One of the passages in the Yazılıkaya limestone quarry contains more than ninety rock-cut reliefs of deities, humans, animals and mythical figures. In 2019, a research team suggested that these were used to record the day of the lunar month, with a stone marker being rolled in front of them on each day, possibly starting from the first day after the new moon. If this is the case, it would make the sanctuary a three-dimensional walk-in calendar.[1]

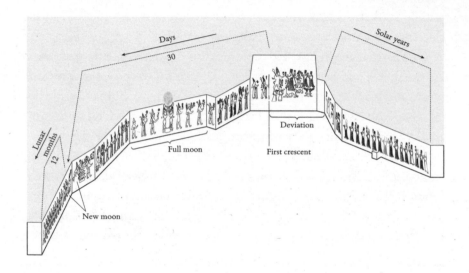

The arrangement of deities in Chamber A.
They are lined up and divided into sections.

In the section of Yazılıkaya now denoted as Chamber A, there are sixty-four deities marking a 30-metre procession. All the deities on the left face north (all male bar two), and those on the right (all female) meet at a climactic plateau housing the supreme deity family. The supreme gods are the storm god, Teshub, also known as the preserver of order in the cosmos, and his wife, goddess Hebat, the chief female deity representing the sun goddess, Arinna. They have one son and two daughters.

There are twelve identically shaped male deities on one side of the sixty-four-deity procession. The research team suggests these could function as counters for the number of synodic months – the time

The deities in Chamber A.

between one new moon and the next – in a year. Next, there are a group of thirty uniformly shaped deities that could have been used to count the days in a month. There are also seventeen similarly sized female deities, but it is thought that two further female deities may have been lost. If there were originally nineteen deities, this could mean that the Hittites were counting the number of solar years before the calendar would need to be adjusted.

On average, a synodic month has 29.53 days, so twelve synodic months would make 354.36 days in a year – short by around eleven days of a solar year. To fill in this gap, the team suggests an extra month could have been inserted at the years 3, 6, 8, 11, 14, 17 and 19. After the nineteenth solar year, there would be a difference of two hours, five minutes and twenty seconds. At the beginning of the twentieth solar year, this deviation could have been adjusted for by resuming counting from the new moon. We may never really know if this is how Yazılıkaya was used, but it's intriguing to think that this sublime monument was a large-scale calendar.

The rise of clocks

For most of human history, measuring time hasn't been about fretting the small stuff. Societies were more interested in developing ways to track the lunar months, planetary motions and the timing of eclipses, believing they influenced their lives, than they were in tracking tiny increments throughout the day. The sun was a more than accurate enough clock for most people most of the time, although that would soon change, with mathematics underpinning the technology that would make it possible.

The gnomon, or shadow stick, was used across the ancient world from about 3500 BCE to estimate the time of day. The slim bar cast a shadow, and its length is then measured. A sundial is a gnomon with a display added, and these were located in public places so people could check the time throughout the day. If accurately made, sundials can tell the time down to the minute, but they have an obvious drawback – they don't work at night, or when it's cloudy.

Enter the water clock, an innovation that took hold in Egypt, Persia, India and China and took the form of a bowl that was slowly either filled with water or emptied. Markings measuring the flow of water were used to measure the flow of time. Clay tablets from around 2000–1600 BCE in Babylonia indicate that water clocks were used to measure how many hours had passed – useful, for example, for a guard at a tower to know how much of their shift remained.

However, one drawback with water clocks was that they weren't very accurate, at least not initially. But then came the elephant clock, created by Ismail al-Jazari, a polymath and inventor from Mesopotamia, in 1206. The elephant clock was a marvel to behold and is one reason why al-Jazari is sometimes now known as the father of robotics. The mechanics were held inside a canopy on top of a model elephant and the clock featured a bird chirping, a serpent and a human-shaped automaton that hit a drum on the half-hour.

Inside the elephant clock was a bowl floating inside a bucket. Water slowly dripped into the bowl over the course of half an hour, making

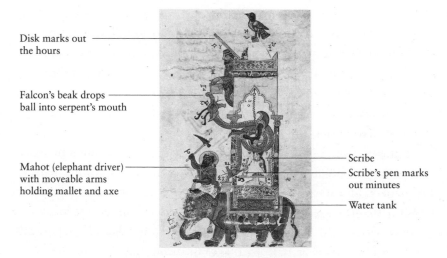

Disk marks out the hours

Falcon's beak drops ball into serpent's mouth

Mahot (elephant driver) with moveable arms holding mallet and axe

Scribe

Scribe's pen marks out minutes

Water tank

Al-Jazari's elephant water-clock.

it heavier, so it sank further into the water, pulling a string attached to the top of the elephant and releasing a ball. The ball then fell into the mouth of the serpent, pushing it forwards and pulling the sunken bowl out of the bucket. Then the bird chirped, a serpent and a human striking the cymbal to mark the half-hour, and the cycle started again. Each element was in mathematical harmony to ensure the whole clock kept time accurately.

The whole set-up was incredibly accurate. However, it did have a drawback – it was massive. You couldn't carry an elephant clock around with you while you went about your day. It was expensive as well, which put it out of reach for most. Al-Jazari himself might have noticed this. He improved other clock mechanisms such as candle clocks and boat clocks, both of which were smaller and simpler.

A more practical time-measuring device was the sandglass, or hourglass. When long-distance sea voyages really took off in the fifteenth century, these became popular with seafarers. Marine sandglasses were made to measure thirty-minute increments. The problem was that sailors had to flip the glasses every half an hour. If they forgot, they'd lose track of the time at their point of origin, which was crucial for determining their position at sea. Flipping duty was one of the most vital – and annoying – jobs on board.

Fully mechanical clocks emerged in the seventeenth century, when polymath Christiaan Huygens and clockmaker Salomon Coster successfully built a wall clock based around a swinging weight in the Netherlands in 1656. It was the world's first pendulum clock. Galileo Galilei had explored the idea of such a clock in 1602, inspired by watching the back-and-forth motion of a lamp swinging in the cathedral of his hometown, Pisa. He noticed that the time period of a swinging pendulum is constant and independent of how far it swings from side to side. This mathematical constancy meant it could be used to record a regular time interval. However, Galileo didn't manage to build a working device himself.

After the Jesuit missionary Matteo Ricci presented a chiming clock that sounded at half past and quarter past the hour to the emperor of Ming China, Wanli, in 1601, successive emperors became

Huygens and Coster created the first pendulum clock.

fascinated with clocks. During the reign of the Kangxi emperor between 1661 and 1722 in the Qing dynasty, more than four thousand clocks and watches made their way to the Forbidden City from Paris and London. Clock manufacturing boomed in China during the Qing dynasty. Each item was expensive, so clocks signified high status, and they were decorated as pieces of art in their own right.

The balance wheel was developed at a similar time to pendulum-based clocks, giving rise to a wealthy class of pocket-watch wearers in Europe. Invented in Germany in 1510, balance wheels rotated back and forth in a similarly mathematically predictable way to pendulums, so they too could be used for watch mechanisms. Pendulum clocks didn't work well at sea because of the movement of the water, but these did.

The term 'pocket watch' was coined by Charles II of England in the seventeenth century and he started the 'high-fashion' trend of English gentlemen wearing them. Pocket watches made great gifts, and once the fashion spread to North America their faces became more elaborate, often embedded with diamonds and jewels, and these could of course be afforded only by the elite. Wristwatches used the same technology and were primarily worn by women. Mass production led to watches becoming cheaper, and timekeeping began to weave itself into day-to-day life.

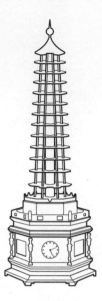

Eighteenth-century models of pagoda clocks were brought into the Qing imperial court. The clocks played music and the tiers of the pagoda moved periodically.

As timekeeping became more a part of everyday life, problems started to arise. In England, for example, coach services took people and mail around the country, adhering to a strict schedule. But as the time was still based on the movements of the sun, such punctuality was difficult. Drivers had to adjust their watch when they arrived in different towns with different time zones, sometimes just minutes apart. And the boom in railway travel in the nineteenth century put the problem on steroids. Over the course of a single journey, a train would stop in a number of different time zones, making it confusing for everyone involved. This led to the idea of all railway companies sticking with the time in London, known as Greenwich Mean Time. Some towns stuck to their own times, but eventually, by the middle of the nineteenth century, the observatory at Greenwich sent the official time via telegraph lines laid alongside the railways for the stations to set their clocks each day. Other countries, like the US, developed similar systems. Eventually, the battle of the clocks would also come to an end in India too. The University of Mumbai senate voted to keep 'Bombay

The White Rabbit with a pocket watch. This illustration is in
Lewis Carroll's *Alice's Adventures in Wonderland*, published in 1865.

time', but after India gained independence from the United Kingdom
in 1947, a universal time was agreed across the country.

About time

As the world slowly got to grips with timekeeping on a global scale,
the understanding of time began to change. In the early twentieth
century, Albert Einstein showed in his special theory of relativity
that time is relative: it isn't a universal constant but changes depending on your location and circumstances.

Einstein's theory described how, if there are two clocks sitting in
spaceships, one of them at rest and the other moving quickly, the

moving clock will tick more slowly from the point of view of a stationary observer. He also showed that the same effect would happen if the moving clock was stationary but nearer to an object with a lot of mass. This time dilation is not easily noticeable, but with our increasing ability to measure time accurately we have not only seen this in action but need to adjust for it every day. Modern navigation systems, such as BeiDou, Galileo and GPS,* use satellites orbiting Earth at a speed of around 14,000 km/h. Measuring the time is crucial to providing accurate positional data and so the satellites carry extremely accurate clocks. But they experience time dilation. This means that GPS clocks, for example, would be out by about 38 microseconds, or 0.000038 seconds, by the end of each Earth day if they weren't adjusted. That's equivalent to the positions they give being off by around 10 kilometres.

Though time dilation is a hard concept to wrap your head around, there is something reassuring in the discovery that time is relative – after all, for most of human history, we've thought of time as relative, using the amount of sun in the day as the basis for our clocks.

Atomic clocks are currently the most accurate timekeeping devices we have, with an error of only around one second in every 100 million years. These use the rhythms of atoms to set the time: a second is defined as the time it takes for a caesium atom to oscillate 9,192,631,770 times. However, our focus on better and more reliable timekeeping conveniently sweeps a fundamental question under the carpet. Yes, it's important to know *what time it is*, but perhaps more important is the question *what is time?*

Another of Einstein's theories, the general theory of relativity, put forward the idea that time is as physical as space. This amalgamation of space and time, known as spacetime, should be viewed as one coherent four-dimensional space. This space can be warped by gravity – and so produces time dilation. The idea of spacetime has been incredibly successful in helping to describe the universe, but it has a strange consequence – the equations involved have no implicit

* The satellite navigational systems of China, the European Union and the US respectively.

direction. This means that time could run backwards or forwards, despite how we experience it.

Key to the puzzle may be the rules that govern heat. The second law of thermodynamics is one of the few physical laws that has some form of directionality; it states that entropy, sometimes thought of as disorder, always increases. In other words, if you leave something for long enough it will always jumble up. This may explain why time always flows forwards (at least it appears that way to us); then again, it may not. Gravity is currently the only force that is not described by quantum physics. As time and gravity are intrinsically linked, this means there is a schism between our understanding of gravity and time on the one hand and our understanding of the quantum world on the other. Many efforts are being made to reconcile these differences by using mathematics to create a quantum theory of gravity. If one is proven, it will almost certainly force a rethink on what we believe time to be.

Fundamental rethinks happen only rarely and when they do they can be so revolutionary that they are difficult to truly grasp. The discovery that the Earth was round must have been one such moment, where the very ground people walked on suddenly seemed different. A little hesitation in accepting the round-world order would have been understandable. Eventually, though, the old orthodoxy is replaced with the new, and what once seemed impenetrable becomes intuitive. Few people today believe the Earth is flat; it just seems so natural to think of the planet as round. Mathematics, too, has been through these earth-shattering moments. The discovery of zero is one of them. We will come to zero's origin story in a moment, but first we need a modern expert on what seems mathematically natural to humans and what doesn't. For that, let us turn to a four-year-old.

5. On the Origin(s) of Zero

One day in the early 2010s, a four-year-old walked into a small room containing a seventeen-inch computer screen and sat down in front of it. On the screen were two squares, each containing either zero, one, two, four or eight dots. This child, like a handful of others who had found themselves in the same situation, had just one task: select the box with the fewest dots.

Psychologists Elizabeth Brannon and Dustin Merritt had set up this simple experiment to test how well four-year-olds understood numbers. It was extremely revealing. When the children had to compare boxes containing one or more dots, they got it right three quarters of the time. But as soon as zero dots was added as an option, the success rate dropped to below 50 per cent – the amount you would expect to get right by chance alone.[1]

There was clearly something about zero that the children hadn't yet got to grips with in the way they had with the other numbers. And perhaps this hints at why humanity took so long to get to grips with zero too.

Zero feels primitive, as if it should have been around since the beginning of time – as if zero is Stone Age technology and the rest of the numerals are post-Industrial Revolution. We've come to think of it as a numerical foundation – how could you build a 'number house' without starting at zero? But this viewpoint couldn't be further from the truth. Not only do we appear to gain an understanding of it later in life, zero is a number that arrives long after the others in human history. And even then, it took mathematicians hundreds of years to really understand it as a number in its own right, rather than just a piece of convenient notation. This development was a momentous event in the history of mathematics, but it was also a murky one with many different possibilities and interpretations.

Though it was the Maya who probably came up with a symbol

for zero first, it was during the so-called Golden Age of India between the fourth and sixth centuries that its importance became fully appreciated. During this period, mathematics leapt to new and exciting heights. A truly fit-for-purpose number system was yet to exist, so mathematicians often altered and played with the numerals they used, and in the course of doing this they invented new ones too. It's fair to say that nothing has a story quite like nothing.

The Golden Age of India

The Golden Age of India began with Chandra Gupta I, who controlled a massive expanse of territory in the northern part of the Indian subcontinent. It is not known exactly how he got hold of so much land, but his marriage to Princess Kumaradevi from the Lichchhavi clan, which may have controlled north Bihar and Nepal, probably helped. His son and subsequent heirs went on to expand the empire through various invasions and battles, all but eliminating as many as twenty-one other rulers in the region.

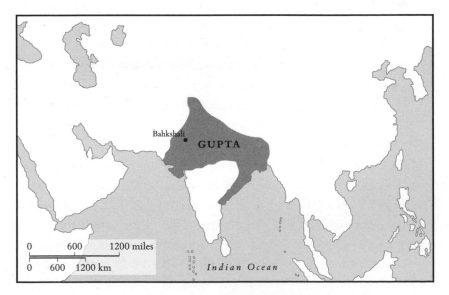

Map of Gupta empire around 450 CE.

Mathematics was valued highly in the Gupta empire. Scholars disseminated knowledge to younger ones through the chanting and recitation of texts. As one ancient verse put it: 'As are crests on the heads of peacocks, as are gems on the heads of snakes, so is mathematics at the top of all branches of knowledge.'[2]

Although there had been intellectual exchanges between India, Egypt, Babylonia and China before this, it was around this time that Indian mathematics and astronomy started to develop in their own right. Astronomy in the Gupta empire was intimately linked to folklore. Astronomers made empirical measurements to determine the lengths of lunar months and the movement of the stars, but this was underpinned by mythology. Eclipses, for instance, were thought to be caused by demons covering the faces of the moon and the sun.

There are few surviving details about women's scholarly role during this period. Some hymns were written by female poets, and some female names, for example, Indrani and Sachi, are listed as authors in ancient Indian texts. Some upper-class women received an education and had the freedom to pursue further study, though they were not considered equal to men and were not allowed to pursue a career. Outside of the upper classes, women were often restricted to the roles of wife and mother and had few legal and property rights. However, despite this, as Gupta women had the freedom to participate in religious rituals and public ceremonies, they did have a greater degree of freedom than during the rule of some earlier dynasties.

The Gupta dynasty actively promoted Hinduism, but they tolerated Buddhism and Jainism too. Priests from this last group were particularly active in the mathematics of the time.

From the fourth or third century BCE, Jain people produced literature that held *sankhyana* – the science of numbers – in particularly high regard. One surviving Jain text, *Tiloyapannatti* by Yativrṣabha, focused on understanding the universe. We know almost nothing about Yativrṣabha's life, but his work mentions some of the names of his teachers and some earlier works on mathematics, which suggests Jain mathematicians passed on knowledge through private tutoring. *Tiloyapannatti* discusses various ways of measuring time and distance;

the ideas in it are based on the Jain belief at that time that there were two suns, two moons and two sets of stars.

For Jain people, there was neither beginning nor end to the universe; space and time were eternal and continuous – the universe had always existed and always would. As a consequence, they became a little bit obsessed with the mathematics of large numbers. They had a period of time, *shirsa prahelika*, which is equivalent to $756 \times 10^{11} \times 8,400,000^{28}$ days – a number that is 208 digits long. Among other important massive numbers was the *rajju*, which is the distance travelled by a god in six months (approximately a million kilometres), and the *palya*, the time it takes to empty a vessel filled with wool by removing one strand every century. This fascination with large numbers led Jain mathematicians to think beyond the ginormous. They started to gather thoughts on infinity which, although mathematically imprecise, were *rajju* ahead of anything else on the planet at the time, or for hundreds of years afterwards.

The Jain people put numbers into three categories: the enumerable (countable), the innumerable (not countable) and the infinite. Enumerable numbers consisted of everything from two to the 'highest number'. The Jain people didn't actually have a highest number, but they used their understanding of massive numbers to get some sense of it. For example, they would imagine filling Earth-sized troughs with mustard seeds and counting them, then declare that this was still not as large as the highest enumerable number. The innumerables were larger than the enumerables but not yet infinite. Jain mathematicians split infinity into five different groups: infinite in one direction; infinite in two directions; infinite in area; infinite everywhere; and infinite perpetually. Understanding and modelling the universe was an important part of Jainism.

The Jain people were the first people we know of to consider that there is more than one size of infinity. This idea would not take root again until late in the nineteenth century, when German logician Georg Cantor worked on the nature of infinity, and even then it took many mathematicians a long time to come to terms with Cantor's findings, and questions about the exact size of different infinities still remain today.

Infinity was not the only area in which the Jain people were particularly proficient. They also carefully examined rules relating to the different ways that objects or things can be combined. For example, they worked out that the six different tastes (bitter, sour, salty, astringent, sweet, hot) could be combined in sixty-three different ways. They explored similar ideas in poetry, producing formulas for the ways in which short sounds and long sounds could be combined to create different metres.

This study of permutations and combinations led them to discover a mathematically important triangle they described as a 'mountain with a peak'. They named it Meru Prastara. The triangle has been discovered by many different cultures and is important in probability theory. Today, it is often called Pascal's triangle, after the seventeenth-century French mathematician Blaise Pascal.

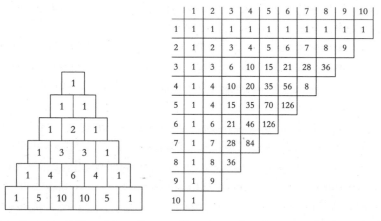

	1	2	3	4	5	6	7	8	9	10
1	1	1	1	1	1	1	1	1	1	1
2	1	2	3	4	5	6	7	8	9	
3	1	3	6	10	15	21	28	36		
4	1	4	10	20	35	56	8			
5	1	4	15	35	70	126				
6	1	6	21	46	126					
7	1	7	28	84						
8	1	8	36							
9	1	9								
10	1									

Meru Prastara (*left*) and Pascal's triangle as drawn by Blaise Pascal in 1665 (*right*). The original text of the Meru Prastara contained only words and did not have this chart; an illustration was added in the tenth century.

The Jain people were also early proponents of exponentiation and its consequences. They wrote, 'the first square root multiplied by the second square root is the cube of the second square root'. Or, in modern notation, $a^{\frac{1}{2}} \times a^{\frac{1}{4}} = (a^{\frac{1}{4}})^3$. The Scottish mathematician, physicist and astronomer John Napier and the Swiss mathematician and maker of clocks and astronomical instruments Jost Bürgi independently

rediscovered these ideas around the beginning of the seventeenth
century.

Why don't we fall off the Earth?

One of the most famous mathematicians of the Gupta dynasty was
Āryabhaṭa I, who lived between 476 and 550 CE. The works he left
behind were extremely influential: many mathematicians copied them,
commented on them and distributed them across India and beyond.

Āryabhaṭa worked at a Buddhist research complex called Nālandā
in Bihar, in the eastern part of the Gupta empire. The complex was a
centre for advanced study, similar to a modern-day university. It
consisted of monasteries, temples and an observatory where people
studied a variety of subjects, including mathematics and astronomy.
The main form of teaching in India up until this point had been
teacher–student pairs, so university teaching was a new development. Students and researchers at Nālandā collected, read and
annotated existing scholarly works, including those created by Jain
mathematicians.

Possibly while at Nālandā, Āryabhaṭa put together *Āryabhaṭiya*, a
treatise on astronomy, cosmology and mathematics that consisted of
around 120 Sanskrit verses. In it, he did not dismiss the idea that
demons caused astronomical events but instead worked them into his
research. When looking at the lunar and solar orbits, he used his findings to explain the movements of the demons involved in eclipses
and tweaked the relevant folklore.

Āryabhaṭa's greatest achievement was to synthesize and interpret
much of the mathematical knowledge that came before him.
Āryabhaṭiya starts with the remark that multiplication by ten promotes a number to 'the next higher place'. Āryabhaṭa was clearly on
to the decimal system, although it was not fully realized in his work.
He then went on to show his workings for finding square and cube
roots, which were written in verse. Although his methods were
incomplete, they provided a foundation for later mathematicians to
build upon.

Āryabhaṭa is often called the father of algebra. In *Āryabhaṭiya*, he wrote that 'if the difference in known quantities of two persons is divided by the difference in their unknown amount, the result shall give the value of the unknown quantity'.[3] It was a vague statement, to be sure, but it did indicate evidence of the very beginnings of algebra. As a later commentary put it, the statement could apply to questions such as the following: there are two farmers, one with a hundred rupees and six cows, the other with sixty rupees and eight cows. If their possessions are worth the same, what is the value of one cow?*

Another reason why Āryabhaṭa is called the father of algebra is his use of quadratic equations. He obtained some solutions to equations of the form $x - y = m$ and $xy = n$, where the task is to find unknowns, x and y, or equations of the form $ax^2 + bx + c = 0$, where the task is to find an unknown, x. He was able to solve these in specific instances.

Like mathematicians in Egypt and China, Āryabhaṭa also approximated pi. He found it to be 3.14159265, which was correct to eight decimal places, though how he managed this is unclear as his methods do not appear in the book. He used pi to prepare a sine table for investigating the features of circles. Measurements of the area and circumference of circles were a necessary part of astronomical calculations.

He also asserted that the motion of the stars in the sky derived from the rotation of the Earth about its axis. He estimated that Earth rotates 1,582,237,500 times per *yuga* cycle, a Hindu time period of 4,320,000 years. This gave Earth's period of rotation as 23 hours, 56 minutes and 4.1 seconds. While this was an amazing insight, and one which we know today is true, he was not the first person to suggest that Earth rotates in this way. Herakleides, a contemporary of Aristotle, had previously suggested it, but it had long been forgotten and ignored. The same fate befell Āryabhaṭa after his death, with many commentaries on his book simply omitting this particular claim. It may seem strange today, but without much concrete evidence it's not hard to understand why astronomers at the time might have had difficulty in

---

* One cow is worth twenty rupees.

believing that it was the Earth and not the sky that rotated. If the Earth really was spinning, how come we don't fall off?*

Parameśvara and Nīlakaṇṭha Somayaji, fourteenth- and fifteenth-century astronomers in India, also supported this view of the solar system, as did Nicolaus Copernicus and Galileo Galilei in Europe a little later. But it wasn't until 1851, when French physicist Léon Foucault demonstrated the effects of the Earth's rotation with a pendulum that the idea was truly cemented.

Āryabhaṭa also invented, or at the very least used, a completely different number system to the others in use at the time. Numerals were often evolving, with more than one number system in play. Many people used Brahmi numerals, which were graphical in nature, like the glyphs used by the Maya. This system had symbols for the numbers 1 to 9, and also for 20, 30, 40, and so on. This meant that zero had no role to play – rather than writing the number 20 as two tens and zero digits, you would use the individual symbol for twenty.

Numerals	1	2	3	4	5	6	7	8	9	10	20	30	...	100
Brahmi	—	=	≡	+	h	Ɛ	ʔ	ち	ʔ	∝	θ	⅃		ʔ

Brahmi numerals.

In the fourth century, Gupta numerals made their appearance. They were more cursive and symmetric, allowing scribes to write them more quickly, and were largely based on the Brahmi numerals. although they did not replace them completely.

Numerals	1	2	3	4	5	6	7	8	9	10	20	30	...	100
Gupta	—	=	≡	ʓ	⅄	⅂	∩	ᛤ	3	℉	⊙	ᴊ		℈

Gupta numerals.

Gupta numerals enabled speed, but they were still cumbersome for writing large numbers in mathematical and astronomical calculations

---

* Thanks, gravity!

(although this hadn't stopped Jain mathematicians from exploring large numbers using Brahmi numerals). So Āryabhaṭa invented his own system, which used the Sanskrit alphabet. He assigned each of the thirty-three consonants of the Sanskrit alphabet to a particular number: 1 to 25, as well as 30, 40, 50, 60, 70, 80, 90 and 100. He then used vowels to express powers of 10. This allowed him to specify large numbers, for example, the number 1,582,237,500 could be written as the word *niśibunḷikhṣhṛi*. Though it's not explicitly used in his work, hidden in it is the suggestion that Āryabhaṭa knew about zero. His own number system seems to be underpinned by how zero fits in, and some of the calculations in *Āryabhaṭīya* are impossible without it.

Āryabhaṭa died in 550 CE, the year that the Gupta empire collapsed, when a nomadic tribe, the Huns, invaded. The empire disintegrated into regional kingdoms and some of the learning centres were disbanded. However, although the new regional kings were not Buddhist themselves, some of them became patrons of Buddhist research complexes such as Nālandā, as well as observatories and associated astronomers after the fall of the Gupta empire.

The numeral with a hole

Probably the most important mathematician from the Gupta era is Brahmagupta, who lived in the seventh century in Bhinmal, in the western part of India. His father, Jishnugupta, was an astronomer and mathematician, which meant that Brahmagupta had access to the work of Āryabhaṭa. At the age of thirty, he wrote his own text, one that ambitiously combined the knowledge of the time with his own work. The book was called *Brahmasphutasiddhanta*, which translates to the *Correctly Established Doctrine of Brahma*.* There is no doubt that the section that had the biggest impact was the one on zero.

Zero had been around for some time before Brahmagupta, with the earliest known example from the Maya civilization, by 300–200 BCE, using a symbol that looked like a shell as a placeholder – one of

* Ahh, the confidence of youth.

the simplest functions of zero. Zero as a placeholder is an incredibly useful shift in how to think about numbers that we still use today.

Recall that when we write the number 201 it is with the implicit understanding that the number furthest to the right represents one unit, then, moving left, there are zero tens and two hundreds. Without a placeholder zero, there would be no way to distinguish between 21, 201, 2001, and so on.

Compare this to Roman numerals, which do not use placeholders in this way. To write the number 201, a Roman would write CCI, where each C stands for 100 and I stands for 1. There's no symbol for zero, as 'tens' simply don't figure in this number. This is fine in some situations, but problematic if you want to carry out a simple calculation such as addition. Adding 99 to 201 is easy: you first add the units, then the tens, and then the hundreds. But try adding CCI to XCIX and it's enough to make you wonder how the Romans managed to achieve very much at all.

The Maya number system was different, using position to great effect and built around powers of 20 rather than units, tens, hundreds, and so on. By about 400–300 BCE, the Babylonians also had a symbol for zero. Initially, it looked more like a colon, but eventually it developed into the tilted cuneiform symbol below.

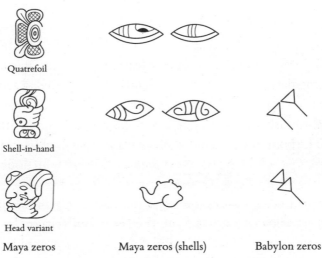

Quatrefoil

Shell-in-hand

Head variant

Maya zeros   Maya zeros (shells)   Babylon zeros

Maya zeros. There were many variations in the patterns on the shell.
Other zeros came from Babylonia, Egypt and Greece.

At some point during the Gupta era, Indian mathematicians began to use a dot to denote the number zero. Brahmagupta certainly used it, so its origin can be traced back to at least the sixth and seventh centuries. But there is also a mysterious document called the Bakhshali manuscript that may have a much older zero. That is, depending on who you ask.

The Bakhshali manuscript is the remains of seventy birch folios featuring arithmetic, elementary algebra and geometry that were discovered by a farmer in Bakhshali (now in Pakistan) in 1881. It is written in a blend of Sanskrit and local dialect, and the author is unknown. Originally, the plan was that the Bakhshali manuscript would be sent to the Lahore Museum, but under British rule it was taken to the Bodleian Library in Oxford, where it has remained ever since.

A few years ago, the Bodleian decided to use radiocarbon dating on the manuscript and found that the three remaining folios come from three separate periods, around 300, 700 and 900 CE.[4] Some argue that this means that Indian mathematicians had zero by the year 300, while others maintain that it was written after 900 CE. The ink itself has not been tested so we have no way to be sure who is correct.

Bakhshali manuscript: Several dots are shown here,
the one on the bottom line functions as a placeholder.

An additional complication to the Indian zero is that similar numerical symbols to those used by Brahmagupta and in the Bakhshali manuscript pop up elsewhere around the same time. There's a dot zero in Cambodia on a stone in the ruined temple at Sambor on the Mekong River that dates to 683 CE. What is now Indonesia has circle zeros from around this time too. The oldest known actual circle zero in India is one that is inscribed at a Hindu Temple in Gwalior that was built around 875 CE.

Khmer zero from 683 CE.

A circle zero found on Banka Island off Sumatra. The number shown is 608, intended to record a year in the Śaka era, a year that would now be dated as 686 CE.

The Sumatra region, including the Srivijaya and Khmer empires, was greatly influenced by both the Indian subcontinent and China.

Gwalior zero in the second line. Around 875 CE.

Islam migrated there from the Middle East, and monks brought Buddhist practices from China. It is possible then that zero was imported to
the islands – but it is equally possible that local scribes developed zero
on their own and the symbol journeyed in the opposite direction.

This makes the exact story of how the modern symbol for zero
came into existence all rather murky. But as far as we can tell, outside
of India, none of these zeros ever went beyond being placeholders.
The first concrete evidence of that change comes from Brahmagupta.
And this was a landmark moment. Zero as a placeholder is a handy
innovation; zero as a true zero is a conceptual leap. In the *Correctly
Established Doctrine of Brahma*, Brahmagupta presented a way to use
zero in arithmetic calculations, making zero a fully-fledged number
that could be added, subtracted, multiplied and divided. He wrote,
'When zero is added to a number or subtracted from a number, the
number remains unchanged; and a number multiplied by zero
becomes zero.' In modern mathematical notation, $x + 0 = x$, $x - 0 = x$
and $x \times 0 = 0$.

He then went on to explain these rules in terms of fortunes (positive numbers) and debts (negative numbers):

A debt minus zero is a debt.
A fortune minus zero is a fortune.
The product of zero multiplied by a debt or a fortune is zero.

He also looked at multiplying by zero (which gives zero), but then
drew this conclusion for division:

Zero divided by zero is zero.

On this point, Brahmagupta was incorrect, but the world would need to wait for the rise of calculus, nearly a millennium later, to find that out.

It's hard to overestimate the importance of this shift from viewing zero simply as a placeholder to it becoming an actual number. Without it, so much of modern mathematics and our understanding of the world would not be possible. The modern world is built on digital technology underpinned by binary mathematics – the mathematics of 0 and 1. Every pixel on your screen, every calculation your computer performs and every datum it stores relies on the impressive power of 0 and 1. Electronically, this comes down to whether something is on or off, but understanding how this works comes down to understanding 0 as a true zero – not just a placeholder. It has to be a zero that can be manipulated and used in calculations.

The question then is why, when so many civilizations seem to have had the concept of zero, it was only in India that it developed further? This may be down more to philosophy than mathematics. While some cultures seem to have more of a fear of or distaste for the idea of nothingness, many in India embraced it. The Sanskrit word *śūnya*, which appears in literature in the early years of the Gupta dynasty, means 'empty' or 'void' and comes from Buddhism. The words for both 'atmosphere' (*ākāsa*) and 'sky' (*kha*) were used to denote zero. Āryabhaṭa, for example, used the word *kha* for the empty position in his numbering system. Others called zero *nirguna Brahman*, meaning 'attributeless truth'. Underlying all these words was the view that nothingness was important, that the infinitude of the world emerges from emptiness. This may have given mathematicians like Brahmagupta the motivation to integrate the concept more deeply than anyone had before.

Zero to hero

Zero didn't spread around the world right away. Its spread started with Arabic-speaking traders who travelled across Africa, Asia and

Europe and embraced the concept, along with an updated version of Gupta numerals, thanks to a book called *Concerning the Hindu Art of Reckoning* written by ninth-century polymath Muhammad ibn Mūsā al-Khwārizmī. Al-Khwārizmī had quickly seen the utility of a number system featuring a true zero, but for many places it was more of a slow burn.

One of the first ways zero arrived in Europe was through the Italian mathematician Leonardo Fibonacci. He had come across the number in the latter part of the twelfth century when he was in the port of Bugia, now Béjaïa in Algeria, supporting his father in representing Italian traders. He probably learned of zero from the Arabic-speaking traders there and by studying al-Khwārizmī's book. When he returned to Europe, Fibonacci published his own book, *The Book of Calculation*, in which he outlined this Indo-Arabic number system, including zero, and how to use it in calculations. He translated the Arabic word for 'empty' in al-Khwārizmī's book – *sifr* – as the Latin *zephyrum*. The Italians in Venice took this word and changed it to 'zero'.

Though Fibonacci championed the mathematics used by Arab traders, he used zero only as a placeholder for counting in the decimal system in his book, so it was not a true zero. Neither did his book lead to the wide-scale adoption of zero. Many European intellectuals were prejudiced against Muslim scholars and so didn't take Arabic mathematics seriously. Zero was banned completely in Florence, as were all other Hindu-Arabic numerals, on the dubious grounds that they were apparently easier to alter for fraudulent purposes. Europe would of course eventually adopt Indo-Arabic numerals in earnest. Their usefulness compared to Roman numerals could only be ignored for so long, although it did take until the Renaissance, in the sixteenth century, for this to happen.

In East Asia there was exposure to zero and Indo-Arabic numerals from Arabic traders and Christian missionaries, however rod numerals were so useful that they were difficult to supersede and remained dominant until around the nineteenth century. The change to Indo-Arabic numerals came, in large part, as a means of making international trade and communication easier.

In other parts of the world, many mathematical traditions were similarly well established. In sub-Saharan Africa, for example, Sankoré Madrasah in the Timbuktu of the Mali empire was one of the major centres for learning. The early-fourteenth-century Malian king, Mansa Musa, known as one of the wealthiest people in history, invested in books brought by Arab traders to the city and hired scholars at Sankoré Madrasah to study them. Many books came to Timbuktu, but mathematics was not studied initially. Base-2 and base-20 systems were far more dominant across the African continent and so base-10 simply didn't hold much appeal.[5] However, that's not to say mathematics wasn't developed in the region. Outside the main learning centres, local traders had many innovative methods for performing quick mental calculations with large numbers, known as *susu* in Ghana; *tontines* in Senegal and Benin; and *esusu* in Nigeria.

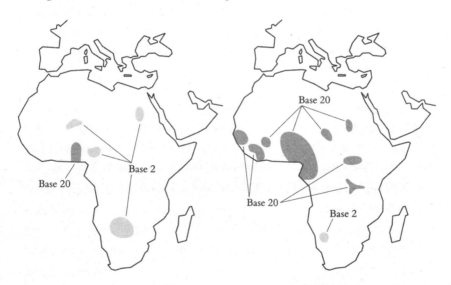

Some of the number systems that have been used in Africa.

In Mesoamerica, the Inca empire, which rose to prominence in the thirteenth century, had its own unique system of recording numbers using a device called a *quipu*. *Quipus* were made from strings of alpaca and llama fleece or cotton, and knots were tied in them to record census figures and tax allocations, as well as names, stories and ideas. When recording numbers, the Inca people used a base-10 system and the knots acted a little like beads in an abacus, essentially making them string spreadsheets. The Inca continued to use *quipus* until 1583, when the Spanish invaders banned them, in the belief that they recorded offerings to non-Christian gods.

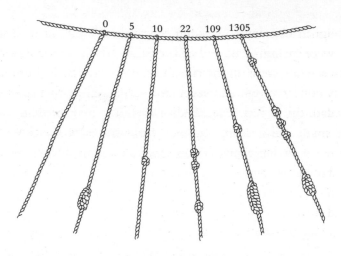

An example of *quipus*.

The first people outside India to appreciate Brahmagupta's work on zero lived on the Arabian peninsula. Here, in the eighth century, a centre of learning quite unlike any other would rise to prominence, and from it would originate some of the most important concepts in mathematics.

6.   The House of Wisdom

In the eighth century something special was brewing in Baghdad. Palaces were springing up to line the banks of the Tigris River. Culture and science were on the rise and there were the beginnings of a scholarly collective whose influence would ripple across the world.

Baghdad, up to this point, hadn't been much more than a collection of small villages, but the new rulers in town – the Abbāsid caliphate – had big plans for an Islamic empire and they wanted Baghdad to be its capital. This was quite the snub to Damascus, which had been the leading city in the region for decades.

It had all begun with the Movement of the Men of the Black Raiment, an uprising against the previously incumbent Umayyad caliphate. The caliphate, at its peak, had ruled from modern-day Portugal in the west to Kyrgyzstan and Pakistan in the east. Perhaps, however, it had always been destined to failure. Though it ruled over a large and diverse population, including people who were Christian, Jewish and Muslim, it was an empire run by Arab caliphs ruling over a largely non-Arab population. This in itself might not have led to the fall of the caliphate; the problem was that people who weren't Arab were treated as inferior citizens. Certain walled cities were accessible only to the ruling Arab class, and non-Arab people were not allowed to hold government posts. This created a deep resentment and alienation among the population and eventually led to revolution.

At the forefront of the revolution was the Abbāsid family. They were also Arab, but they claimed to directly descend from the uncle of the Prophet Mohammed and courted the support of non-Arab people, particularly in Persia. With discontent spreading, disparate uprisings fused together under the banner of a black flag – a symbol of protest against the tyranny of the regime. In Khurasan, a military city of eastern Iran, Abu al-'Abbas as-Saffah led a coup that ousted the Umayyad dynasty, thanks in part to the support of local residents. He became the first caliph of the Abbāsid dynasty in around 750 CE.

As-Saffah died from smallpox after only a few years in charge, and his half-brother al-Mansūr took over and began the process of moving the capital city from Damascus to Baghdad. This appeased the calls from the Persians who had helped overthrow the previous regime to reduce Arab influence in the Islamic empire. It also created a booming cosmopolitan and commercial hub. Baghdad would ultimately become the largest city in the world, home to over a million people.

With this more inclusive society came a hunger and a desire to pursue knowledge for its own sake. An intellectual powerhouse was born and, within it, some of the most important mathematical concepts in history were pondered and proliferated.

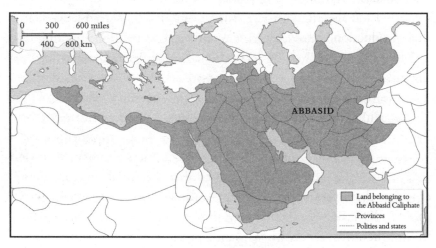

The Abbāsid caliphate in 849 CE.

Calculations for the caliph

To try to consolidate his influence in the region, al-Mansūr embarked on a scientific and cultural mission to translate hundreds of books into Arabic, the official language of the Abbāsid caliphate. Important texts on philosophy, astronomy and astrology were written in Greek,

Syriac, Sanskrit and the Persian language of Pahlavi. By unifying them all together in Arabic, al-Mansūr believed he could also unify the territories ruled by his dynasty. Knowledge and power went hand in hand.

In part, al-Mansūr's drive for these translations stemmed from his personal interests. He was obsessed with astrology and paid for many translations of ancient myths from Persia, despite Islam viewing the subject less than favourably. These ancient Persian myths featured plenty of mathematics and astronomy, despite being scientifically shaky overall. The prominence of the subjects in the stories helped to instigate a more widespread movement towards scientific translation and a thirst for new books and ideas. Around 770 CE, the Abbāsid court discovered the works of Āryabhaṭa and Brahmagupta and promptly translated them into Persian and Arabic. From Greece, the mathematical works of Archimedes, Apollonius, Diophantus, Euclid and Ptolemy were imported and also translated. Al-Mansūr funded these translations, and their production was aided by the recent establishment of paper mills.

The first of these was built in Samarkand (now in Uzbekistan), on the Silk Road between Tang China and the West, after the Abbāsids defeated the Chinese army there and took control of the territory in 751 CE. China already had paper mills, and among the prisoners taken by the Abbāsids were people with papermaking knowledge. The Abbāsids and their prisoners constructed the new Samarkand paper mill and used locally produced raw materials such as flax and hemp crops in their production of paper. They then built several more paper mills in Baghdad. Suddenly, producing books was much easier. Before the mills, records had been kept on perishable products such as plant leaves and bamboo strips; now, with cheaper, sturdier materials at hand, scribes got to work. More books meant more scholars, as copies were circulated to remote places. The age of papyrus and parchment was over. Paper was here to stay.

Successive caliphs continued to collect books and build impressive libraries, but al-Mansūr's great-grandson al-Ma'mūn took it to a whole new level. Al-Ma'mūn was in charge of the now massive Abbāsid army and led it as far as Constantinople in order to expand the dynasty's territory. His father, al-Rashīd, had established diplomatic ties with Chinese and European emperors and increased trade links between the regions. The Abbāsid people specialized in producing silk and rock crystal. There was such demand for this in Europe that the Emperor of the Romans, Charlemagne, was forced to compensate al-Ma'mūn for the overall trade deficit.

As Islam spread under the Abbāsid dynasty, there was less tension between people of different faiths. Christian and Jewish scholars could interact more freely, as early Islam was open to other faiths. Skilled Christian and Jewish translators travelled to Baghdad from within and beyond the Abbāsid borders, knowing that they would be paid well and supported by the caliph. Al-Ma'mūn quickly gained a reputation as a cultured caliph who encouraged original thinking and free debate. He started an ambitious project to collect all the world's books under one roof. This location would become known as the House of Wisdom, and the brightest minds within the Abbāsid dynasty and elsewhere would congregate and be employed there.

The House of Wisdom had a magnificent library – almost certainly the largest repository of books in the world at that time – but it was more than just a library, with scholars engaged in collecting, cataloguing, translating, copying, studying and writing. Thanks to generous grants from al-Ma'mūn, scholars were able to spend all their time in the pursuit of knowledge. They translated works of mathematics, philosophy, medicine, astronomy and optics – the study of light – into Arabic and shared them with each other and across the empire.

One of the biggest challenges was trying to comprehend this vast mass of knowledge and make it consistent. The measurements in the books were all over the place: expressed in different units and often contradicting each other. The House of Wisdom became the place where the world's knowledge was placed under close scrutiny with the aim of working out what was true and what wasn't. When necessary, the scholars performed their own observations, for example tracing the movements of the stars.

Books and scholars at an Abbāsid library.
Illustration by Yahyá al-Wasiti, 1237 CE.

Algebra, Algorithm, Al-Khwārizmī

Of all the scholars who spent time at the House of Wisdom, Muḥammad ibn Mūsā al-Khwārizmī (780–850 CE)* may have had the biggest impact on the world. It's fair to say that, without him, both mathematics and computer science would look very different today.

Many of the biographical details about his life have been lost or are contradictory. He was probably born into a Persian family in the region of Khwarazm – his name literally means 'the native of Khwarazm' – which sits on the border of modern-day Turkmenistan and Uzbekistan but was then part of the Abbāsid empire. His name also gives some evidence that he might have been a follower of one of

---

* His name has many variations in spelling, among them Abū ʿAbdallāh Muḥammad ibn Mūsā al-Khwārizmī and Abū Jaʿfar Muḥammad ibn Mūsā al-Khwārizmī.

the world's oldest religions, Zoroastrianism. But then again, a preface to one of his books suggests he was an orthodox Muslim.

Al-Khwārizmī was employed by caliph al-Ma'mūn at the House of Wisdom. Initially, he gained a reputation as a capable geographer. He prepared a map with hundreds of cities and their coordinates, commonly known as the *Picture of the Earth*, in 833 CE. He probably travelled to many cities and did his own research, but the original list of latitudes and longitudes came from Ptolemy's *Geography*, from the second century. Al-Khwārizmī transferred the coordinates of over 2,400 cities, mountains, rivers and coastlines on to his map. Though it no longer survives, al-Khwārizmī's map expanded on Ptolemy's to span from the Atlantic to the Indian Ocean. But it was in mathematics that al-Khwārizmī truly excelled.

One of the oldest surviving paper maps. A copy of al-Khwārizmī's *Picture of the Earth* made around 1036 CE, showing part of the Nile.

His text on arithmetic, *Concerning the Hindu Art of Reckoning*, took in the work of Brahmagupta, using and translating his decimal system into Arabic. Through this translation, the Brahmi numerals became a new set of Arabic symbols closely associated with the Arabic alphabet. Al-Khwārizmī appreciated Brahmagupta's view of zero and the ease of use of his decimal system. Though al-Khwārizmī's original text has now been lost, it was this text that was eventually disseminated around the world and gave us the number system we use today.

The text originally travelled via al-Andalus, a Muslim-ruled region of the Iberian peninsula. Due to its proximity to Europe, works originating from Greece had long been in the area, but manuscripts from the House of Wisdom started to make their way there too. Toledo, in what is now Spain, would eventually become one of the largest centres of learning and translation in Europe. Arabic mathematics was translated into Old Spanish, Latin Hebrew and Judaeo-Spanish.

Translating al-Khwārizmī's work into Latin led to several ways of writing Indo-Arabic numerals, but it was the style that became dominant in the Western Mediterranean region that would evolve into the modern-day numerals we know today.

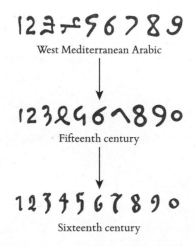

West Mediterranean Arabic

Fifteenth century

Sixteenth century

The Indo-Arabic numerals changed over time and
they became our modern-day numerals.

Despite their eventual spread across the world, decimals were not used by most mathematicians at the House of Wisdom. Many scholars preferred the Babylonian sexagesimal (base-60) system, especially for astronomy, as working with the 360 degrees of a circular orbit was particularly easy. Calculating with it was known as astronomer's arithmetic.

You might think introducing the world to the decimal number system would be enough for one person, but it's just one of al-Khwārizmī's many achievements. His *Compendious Book on Calculation by Completion and Balancing* eventually became the dominant mathematical text taught throughout the Middle East and Europe, and introduced two more of the most important concepts in all of science – algebra and algorithms.

The word 'algebra' comes from the word *al-jabr* in the title. Algebra now refers to the mathematical area of study of unknown values, but the Arabic word originally referred to a particular technique called 'completion' that was used for rearranging equations to help solve them.

Throughout the book, al-Khwārizmī's focus is on solving real-life practical problems, such as those linked to inheritance, land partitions, lawsuits, trade and the digging of canals. One of the key mathematical ideas he developed was a method for solving linear and quadratic equations. In modern notation, these are equations that feature multiples of x in the linear case and multiples of x^2 in the quadratic case. (The equation $2x = 4$ is a linear one; $3x^2 + 2x = 1$ a quadratic one.)

The way al-Khwārizmī solved these problems was by first reducing them to one of the following six forms:

Squares equal roots $(ax^2 = bx)$
Squares equal number $(ax^2 = c)$
Roots equal number $(bx = c)$
Squares and roots equal number $(ax^2 + bx = c)$
Squares and number equal roots $(ax^2 + c = bx)$
Roots and number equal squares $(bx + c = ax^2)$

. . . where a, b and c are positive integers.

Given any linear or quadratic equation, he used the techniques of *al-jabr* (completion) and *al-muqabla* (balance) to reduce them to one of these six forms. With *al-jabr* you could remove negative quantities by adding the same quantity to each side. For example, $3x^2 = 40x - x^2$ could be rewritten as $3x^2 + x^2 = 40x - x^2 + x^2$ which reduces to $4x^2 = 40x$, the first of the six forms. *Al-muqabla* could be used to place the quantities of the same type to the same side of the equation. For example, $3x^2 + 20 = 40x + 5$ could be rewritten as $3x^2 + 15 = 40x$ (the fifth form).

Al-Khwārizmī would not have used this notation, of course – the symbol x wasn't yet used for unknowns, and neither was the equals sign – instead, he wrote everything out in words.

From these six standard forms, al-Khwārizmī provided procedures that people could follow to solve the equations to find out the unknown quantities. Before him, mathematicians such as Euclid, Diophantus, Āryabhaṭa and Brahmagupta had worked on problems involving unknown quantities, but al-Khwārizmī's book showed how to find solutions in a more systematic way. This procedural way of solving problems was a crowning achievement and is the origin of our concept of an algorithm. Although he did not define what algorithms are, these simple step-by-step instructions meant his book gained widespread attention, as it was read, studied and translated further across Europe. The word itself came from the Latin word *algorismus*, derived from al-Khwārizmī, and the Greek word *arithmos* meaning 'number'.

What is an algorithm?

Al-Khwārizmī's idea would become fundamental to modern life. Barely a day goes by without a news headline featuring the word 'algorithm', indicating how they have become ever dominant in our everyday lives. Everything from washing machines to the recommendation engine behind your favourite online shop uses an algorithm. They are so ubiquitous that it's worth taking a moment to understand exactly what they are.

The concept is rather simple. An algorithm is just a list of instructions that can be used to solve a problem or perform a task. A recipe, for instance, is a sort of algorithm. Given the ingredients, anyone should be able to understand how to combine them to make the final dish. A simple algorithm for buttered toast might be something like the following:

Put a slice of bread in the toaster.
Wait two minutes.
Remove the bread from the toaster.
Put butter on the bread.

Most humans could easily follow those instructions, but the key to an algorithm is that the instructions should be unambiguous. A robot could mess up this recipe but still justifiably say that it had followed the instructions accurately. Read through that recipe again, and one can imagine how a robot might end up with an untoasted piece of bread with a packet of butter placed on top. Nowhere does it explicitly say to turn on the toaster and what does *put butter on the bread* even mean?

Writing a good algorithm requires precision. Here's an example of a better one, from Euclid's *Elements*. Euclid didn't quite give out the general formulation as set out below, but he worked out many of the important parts that form the basis of the algorithm.

The point of 'Euclid's algorithm' is to find the greatest common divisor (GCD) of two numbers – the largest whole number that wholly divides them both. For example, the GCD of 9 and 6 is 3, because both 9 and 6 can be divided by 3 and there is no larger whole number that does this. This sort of calculation is helpful when trying to work out, say, the sum of two fractions. But calculating the GCD can be tricky. That's where Euclid's algorithm comes in.

Suppose you have two numbers, A and B, where A is larger than B. Let's write the GCD of them as GCD (A,B). Here's the algorithm:

1. If $A = 0$, then $GCD(A,B) = B$
2. If $B = 0$, then $GCD(A,B) = A$
3. If neither of the above holds, divide A by B and calculate the remainder. Write the number of times B divides A as Q and the remainder as R, so that you have $A = Q \times B + R$.
4. Find $GCD(B, R)$

This may look a little complicated, so let's work through it using 9 and 6.

First, look at condition 1. Clearly, $A = 0$ isn't true, so we can go to step 2. Clearly, $B = 0$ isn't true either, so we can go straight to step 3. For step 3, dividing 9 by 6 and calculating the remainder gives us $9 = 1 \times 6 + 3$. Step 4 says to now calculate $GCD\,(6,3)$.

Again, clearly, we're not in steps 1 and 2 territory yet, so let's divide 6 by 3 to get $6 = 2 \times 3 + 0$. Step 4 then says we must find $GCD\,(3,0)$.

So we go to the top of the list of instructions once more, but now we find that step 2 applies. In other words, $GCD\,(3,0) = 3$.

Nifty, eh?

Of course, in this case, we already knew the answer and it was fairly easy to work out by eye. But if the numbers had been hundreds of digits long, it would have been much harder, although, these days, a computer could have been programmed to follow Euclid's algorithm and calculate the answer on our behalf.

And there lies the beauty of the approach that al-Khwārizmī hit upon at the House of Wisdom. By breaking down a problem into simple steps, the algorithmic approach makes it possible for anyone or anything to solve not just easy problems but very complicated ones too. When you ask your smartphone to plot the best route from Brussels to Bangkok using public transport but not certain metro lines, the same principles are at play – just as they are with many other situations, most of which we never need to worry about, because an algorithm does it instead.

The celestial revolution

The House of Wisdom also revolutionized how mathematical ideas were exchanged. By examining foreign books, translating them and advancing the ideas they contained, the House revamped the way concepts moved between countries, marking the end of ancient traditional mathematics and the dawning of a new era. This shift meant that people started to view mathematical knowledge as something that could be shared between cultures. As the ninth-century polymath Abū Yūsuf Yaʿqūb ibn 'Isḥāq al-Kindī wrote: 'We should not be ashamed to recognise truth and assimilate it, from whatever quarter it may reach us, even though it may come from earlier generations and foreign people.'[1]

Attempting to correct the findings in previous works became a large part of the House of Wisdom's ethos. In translated astronomical texts, the data on the positions of the sun, moon and planets varied greatly. Caliph al-Ma'mūn threw more resources into building observatories so that his astronomers could make observations of their own. The book that especially interested al-Ma'mun was Ptolemy's *Almagest*. It laid out a dominant view of the solar system at the time, with Earth at the centre and the other planetary bodies orbiting it. The Ptolemaic system leaned heavily on the idea that the universe should in some way be mathematically perfect, and Greek astronomers viewed spheres in this light. As such, many astronomers believed that the moon orbited on a sphere that was closest to Earth. Mercury, Venus and the sun then orbited on larger spheres, followed by Mars, Jupiter and Saturn on larger ones still.

Unfortunately, this perfect view of the solar system didn't match reality. Astronomers saw that the movements of many of the celestial bodies didn't follow those predicted by spheres alone, and so Ptolemy had to come up with various modifications, such as secondary circular orbits within the spheres known as epicycles. Though Earth was in the middle, many of the orbits had different points as their centre. The result was an impressively close match to the data they had, even if it did turn out to be based on the wrong explanations.

Arab scholars identified and attempted to fix problems with this model. The Banū Mūsā brothers, three contemporaries of al-Khwārizmī, doubted the length of the solar year as calculated by Ptolemy and set out to do their own research based on their knowledge of Greek astronomy. Al-Ṣābi' Thābit ibn Qurrah al-Ḥarrānī, who worked with the brothers, made the breakthrough, by calculating that the sun's orbit around the Earth took 365 days, 6 hours, 9 minutes and 12 seconds. Although, of course, the sun does not orbit the Earth, this calculation for the length of a year was off by just two seconds.

This constantly questioning approach of Ptolemy wouldn't quite lead Arabic scholars to the right model of the solar system, but it did open up this ancient view for discussion. It would take many years and the collapse of the Abbāsid dynasty before it was overhauled, with religious views of Earth and the heavens playing a large part in keeping it in place.

Burning down the House

The Abbāsid dynasty couldn't last for ever. It grew to such an extent that those in power took for granted the multiethnic and multi-religious foundations on which it was built. Non-Arab groups were once again alienated in the region. New empires emerged from within, fracturing the Abbāsid caliphate.

In 946 CE, Baghdad became the battlefield of two invading forces: the Buyid Emirate of Iraq and the Hamdanid Emirate of Mosul. After several months of fighting, the Buyid eventually took Baghdad and tipped the city into chaos. After a century of instability, the Abbāsid caliph Abu Mansur al-Faḍl ibn Ahmad al-Mustazhir managed to take back Baghdad in the twelfth century, but the Abbāsid dynasty would never again prosper as it had in its intellectual heyday.

The final blow came when the Mongol empire conquered Baghdad in 1258 CE. The invading forces burned down the House of Wisdom, threw books into the river and killed many scholars. The last Abbāsid caliph was executed and the dynasty's days were over. However, this did not mean that all the knowledge it had accumulated was lost. Scholars left Baghdad and created new centres of learning in Diyarbakir (in current-day Turkey), Isfahan (current-day Iran), Damascus and Cairo. Many copies of the books produced at the House of Wisdom had already been sent to major cities such as these in the Islamic empire.

And despite the tragic loss of the House itself, under Mongol rule knowledge continued to proliferate. Thirteenth-century Persian polymath Nasir al-Din al-Tusi convinced a grandson of Genghis Khan's, Hulegu Khan, to build an observatory in Maragha, in modern-day Iran. Al-Tusi questioned Ptolemy's model of the universe and, building on the work that had survived from Baghdad, he used the observatory to interrogate the astronomical data himself. The Maragha Observatory was built in 1259 CE and became a place for a new generation of astronomers, mathematicians, engineers and librarians to congregate.

It was here that al-Tusi would hit upon a fundamental idea now known as the Tusi couple, a new mathematical method for describing the motion of celestial bodies as circles rolling inside bigger circles. The paths traced by a line connected to the centre of a circle turn out to look a lot like the paths of the secondary circular orbits and the epicycles of the moon, the sun and other planets in the sky. Perhaps this underlying mathematics would more closely match reality?

 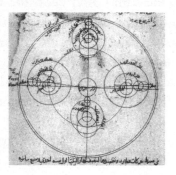

Al-Tusi's drawing of Tusi couples (*left*). Ibn al-Shatir's sketch of a new model (*right*), which centred Earth but drew multiple epicycles. Here, the idea of Tusi couples was also incorporated.

Al-Tusi developed the germ of the idea, but it was Damascus-based mathematician and astronomer Ibn al-Shatir who really made the most of it in the fourteenth century. Al-Shatir valued empirical data above all else and managed to use Tusi couples to build a model of the solar system that matched observations more closely than any model before it. He did not go so far as to put the sun at the centre but, mathematically, he worked out all the kinks to make this possible. This same mathematical framework would underpin Polish astronomer Nicolaus Copernicus's work a hundred years later when he made the seemingly impossible conceptual leap of putting the sun at the centre at the height of the European Renaissance – a time when seemingly impossible leaps became all the rage.

7.   The Impossible Dream

There are few grander ambitions than wishing to be able to fly. When you see a bird leap into the air, flap its wings and fly into the distance it's easy to see why so many *Homines sapientes* have dreamed about doing the same. And Leonardo da Vinci was no different. From a young age, he was fascinated with the idea of flying. As a boy, he drew sketches of bird wings and studied how birds were able to control themselves in the air. As he grew older, this hobby developed into a full-on obsession. He sketched designs for many machines that he believed might one day help humanity to fly.

 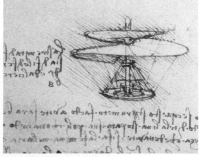

Sketches from Leonardo da Vinci's notebook. Studies of birds (*left*) and an aerial screw to lift the object suspended beneath it upwards (*right*) (*c.*1489 CE).

The technology needed for flying machines was nowhere near ready. It would take three hundred years before the hot-air balloon would first give humans a taste of flight and at least another hundred for planes to take off.

But perhaps his bold ambitions were a symptom of da Vinci's era. In the fifteenth century in Renaissance Europe, there were big ideas aplenty, many of them rediscovered from classical Greek texts during scholarly activities in Italy. Euclid's *Elements* was translated and distributed more widely, establishing its status as the basic textbook of geometry. The works of Pappus and Apollonius were also rediscovered, commented on, copied and amended. Similarly to what had happened in the House of Wisdom, old ideas were respected but also interrogated and corrected. It was around this time that Copernicus demonstrated that the planets revolved around the sun and not the Earth, overturning Ptolemy's view from a thousand years earlier that everything revolved around us.

The trigger for this explosion of ideas came with the fall of Constantinople in 1453. When the Ottoman empire took over the city, many scholars fled, taking their Greek and Latin books with them. The Medici family encouraged them to resettle in Florence, where they bankrolled the study of a broad range of subjects, including poetry, grammar, history, rhetoric and philosophy.

The Renaissance was certainly a monumental moment; historians

The first public demonstration of a hot-air balloon, in Annonay, France, in 1783.

once marked it as the start of the 'Scientific Revolution'. Today, however, the idea of a scientific revolution has come under scrutiny. As we've seen, science had already been advancing at pace for many years across the world. Many ideas that turn up in the 'Scientific Revolution' had already been explored elsewhere or were the culmination of incremental steps made by others. So it seems wrong to call it a revolution or to use 'the' to imply it was the only one. Still, the mathematics developed in and around Renaissance Europe was distinct from what came before. Mathematicians started a habit of

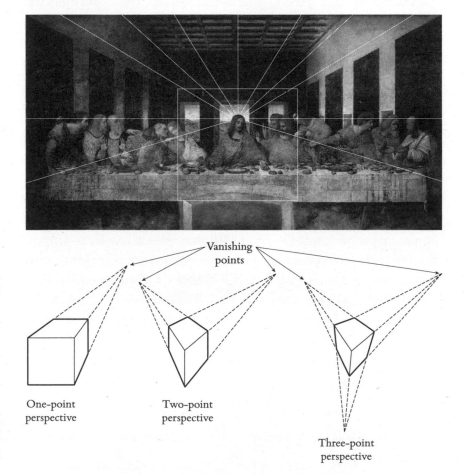

Da Vinci's *Last Supper* with auxiliary lines showing a vanishing point. This is the starting point of projective geometry, whereby mathematicians began to examine geometric figures under projection from a point.

writing letters to each other about the problems of the day. This meant that many minds could make light work of mathematical challenges, making the impossible start to seem less impossible.

A gambler's ambition

Predicting the future was one task that had been considered impossible by most. And then along came Antoine Gombaud. Born in 1607, Gombaud was a dice-playing gambler, writer and philosopher. He was not a nobleman and disliked hereditary power, but when engaging in intellectual debates he would go by the noble pseudonym Chevalier de Méré. He was an adviser to Louis XIV and regularly met with the upper echelons of society, including intellectuals such as Françoise d'Aubigné (sometimes known as Marquise de Maintenon), who would also become both an adviser and a secret wife to Louis XIV. Gombaud believed that the issues of the day should be solved through rational discussion between rational individuals. At salons, he and his friends would discuss topics of importance – or at least, topics of importance to him.

One, which became known as the Problem of the Unfinished Game, began to obsess Gombaud. It goes a little like this: suppose that you and a friend are playing a simple game that consists of twenty rounds, and you each have an equal chance of winning each round. An example might be rolling a die: if it lands on an even number, you win; if it's an odd number, your friend wins. Gombaud's question is: who should win if the game is interrupted partway through?

One option was to simply divide the pot according to the number of wins acquired so far. For instance, if the score was 7–5, then the person in the lead would get $\frac{7}{12}$ of the pot and the other would take $\frac{5}{12}$. However, this seemed unfair, as it was still possible that the person losing would have ended up winning the game if it continued.

And what if the game stopped when the score was 1–0? Giving all the money to the person who was 1–0 up seemed particularly unfair when only one round had been played. Returning to the game later on was simply not an option for these intrepid dice-players. They wanted a winner immediately and so needed some way to predict the likelihood of different future outcomes. Many people thought that it was impossible to predict future outcomes or events, or possible only for God. Gombaud, however, was unhappy with leaving the outcome of the game to the divine and so turned to the best French mathematicians of the time for help.

Blaise Pascal was thirty-one years old in 1654 and already a renowned intellectual in France. When he was a teenager, he had invented a mechanical desk calculator he called the Pascaline to help his father in his job as a tax supervisor. The Pascaline used a sequence of gears to allow two numbers to be mechanically added and subtracted at speed. It was an immediate sensation – the first such calculator to be widely used.

The Pascaline, 1652 version. Pascal made the first version in 1642 and
improved on the design several times.

By the time Gombaud approached him, Pascal was at the zenith of his intellectual power and was focusing on mathematics. He had recently written a treatise on geometry which included his approach to projective geometry – a subject directly linked to the perspective paintings Renaissance artists were producing at the time.

Pascal pondered the Problem of the Unfinished Game and came

up with a possible solution in the spring of 1654. To understand it, imagine two players, Adil and Bao, with two rounds left in an eleven-round game. The score is 4–5, so Adil needs two more wins to be triumphant and Bao needs one. This means that the possible outcomes of the game are:

Adil wins, Adil wins → Adil wins the pot
Adil wins, Bao wins → Bao wins the pot
Bao wins, Adil wins → Bao wins the pot
Bao wins, Bao wins → Bao wins the pot

In three out of the four cases, Bao wins, so Pascal suggested that in this situation she should take three quarters of the money and Adil the rest.

By working out the possible outcomes of any situation, Pascal could, in a particular sense, predict the future. But when mathematician and founding member of the French Academy Gilles Personne de Roberval saw the calculation, he wasn't convinced. He thought that there should only be one case in which Bao wins first, as this would be the end of the game. In his eyes, there were only three outcomes:

Adil wins, Adil wins → Adil wins the pot
Adil wins, Bao wins → Bao wins the pot
Bao wins and the game ends → Bao wins the pot

By Roberval's calculation, Bao should take two thirds of the money. Discouraged, Pascal wrote to his friend Pierre de Fermat. Fermat was a lawyer and mathematician based in Toulouse and a lot older than Pascal. Fermat assured Pascal that his method was the correct one. The probability of any player winning a round is $\frac{1}{2}$. In the example above, this must happen two times in a row for Adil to win overall; in other words, there is a $\frac{1}{2} \times \frac{1}{2} = \frac{1}{4}$ chance of Adil winning. There are only two players, so in Roberval's view this would mean the chance of Bao winning must be $\frac{3}{4}$ rather than $\frac{2}{3}$.

Much of this may seem like simple school mathematics, but these calculations were among the first to put hard numbers on the chances of something happening in the future.

Investigating probabilities had already been attempted by some Italian mathematicians, most notably Girolamo Cardano, who wrote a practical guide for gamblers and was a compulsive gambler himself. His *Book on Games of Chance* from around 1564 CE, although not published until a century later, looked at the odds involved in various dice games, as well as some cunning methods of cheating. He pinned down some of the basic ideas of probability but didn't go much further. Gombaud, Fermat, Roberval and Pascal, however, did. They started to look at more general situations, for example when more players were involved, and tried to consider which rules applied.

The shift in thinking would lead to an entire field of mathematics that aims to assess the likelihood of what is to come and is today used in everything from business calculations about expected profits to the likelihood of a new disease spreading called Probability Theory.

Something to write about

The story also illustrates one of the main ways that mathematics progressed in seventeenth-century Europe: by mail. One person would be tackling a problem and would write down their thoughts and send them to another. Someone else would then weigh in, bringing another perspective or approach. Each letter inched the process forward until a solution was found. Letters were not kept secret. Instead, the academic discussions written in them were meant to be read and shared with others. Throughout the late seventeenth and early eighteenth centuries, intellectual communities in Europe and the Americas thrived by building rich scholarly networks. This movement was later dubbed the Republic of Letters.

Europe started to build learned societies with the support of royal patrons. The Royal Society of London was founded in 1663, and top intellectuals would gather there with the purpose of conducting experiments to help develop and spread understanding.

The French Academy of Sciences opened in Paris in 1666. Its members would meet twice a week and publish scholarly works on mathematics, physics, chemistry and biology. The French Academy gained a reputation for academic authority. Other learned societies sprung up in other major cities in Europe, such as Berlin, Bologna and Rome.

Members of the French Academy of Sciences greeting Louis XIV in 1667, by royal painter Henri Testelin. In the background, you can see the new Paris Observatory.

These societies again shifted the way in which mathematics in Europe was done. Letter exchanges continued, but societies now stamped a seal of approval on the latest proofs and theorems. In an early form of peer review, scholars would discuss and occasionally reject the new findings before the societies published the results. It was the beginning of the professionalization of mathematics and science – albeit for only one half of the population.

To become a member of one of these societies, you generally had to be elected by the members. Even though women weren't explicitly forbidden from entering, in practice, that's what happened for hundreds of years. It wasn't until 1945 that a woman became a member of the Royal Society of London, when crystallographer Kathleen Lonsdale was elected. It took the French Academy of Science until 1979 to elect a woman, mathematician Yvonne Choquet-Bruhat.

However, by the mid-seventeenth century the rise of salons was making it easier for some women to participate in the world of mathematics. Salons were gatherings of intellectuals, often held at someone's home. They would focus on particular themes of interest, with the goal of increasing knowledge about the subject through conversation and the unimpeded exchange of ideas. Women often participated as hosts and attendees. The idea originated in Italy, but it took off particularly in Paris, where wealthy women would show off their collections of curios, such as rare books, clocks and scientific instruments. Having a natural history cabinet was very much in vogue. Among the aristocracy, this was a setting in which men and women spoke together as relative equals, giving women the opportunity to pursue their academic interests and participate in scientific discussions. A similar dynamic also occurred in the Renaissance courts. Here royal men and women mixed, seated alternately where possible, though there were often more men than women.

Compared to modern standards, these were baby steps, but the result was a new generation of aristocratic women who were well educated and able to participate in scientific discussions. This participation would contribute to dismantling the idea prevalent at the time that women were simply incapable of understanding mathematics. In these early years, these women had a huge impact in recognizing and facilitating some of the most important mathematical discoveries of the day.

Exiled but not out

Princess Elisabeth of Bohemia was born in 1618, the year the Thirty Years War broke out. It had started as a revolt, when Protestants in Habsburg Austria expelled the Catholic king and Elisabeth's father, Frederick V, was elected to rule the kingdom of Bohemia. Along with Elisabeth's mother, Elizabeth Stuart, a daughter of James I of England (James VI of Scotland), Frederick moved to Prague in August 1620. This was meant to signal a new era of political stability, merging English and continental Protestantism.

Frederick's reign was initially supported by his fellow nobles, but they formed a new alliance after finding out that James I of England did not approve of his son-in-law's acceptance of the Bohemia crown and would not be offering his support. Knowing that help and resources would not be forthcoming from James, Frederick's alliances fell apart and he lost the throne just over a year later, his short reign earning him the sobriquet the 'Winter King'.

He had to flee, and Elisabeth had to go into hiding. When Frederick finally settled in The Hague (then under Spanish rule, now in the Netherlands), he was able to summon the rest of his family to reunite in his newly established 'Exile Court'. Politics descended into chaos and the region into war. Initially, only the southern part of Germany was affected, but the conflict developed into a full-scale war between Denmark and Sweden, which eventually spread into northern Germany. The Netherlands and France soon joined as Habsburg hegemony collapsed.

Despite this tumultuous backdrop, Elisabeth, as a royal, was protected, and she received a good education from her personal tutors. Her siblings called her 'La Grecque' for her mastery of Greek. Her knowledge of philosophy was outstanding, and she also studied painting, music, dancing, Latin, French, English and German – and mathematics, which she learned from Jan Stampioen, a tutor of Prince William of Orange, and the polymath Christiaan Huygens. Elisabeth loved learning new things, so much so that she even taught herself to perform dissections on small animals.

The Hague was a good place for a royal hungry for knowledge. The top intellects of Europe gathered there, and at the age of sixteen Elisabeth began hosting debates in the Exile Court. It was here that she first met René Descartes.

Descartes was born in central France in 1596 and in adulthood he travelled around mingling with the courts and the military, wishing to be 'mixing with people of diverse temperaments and ranks, gathering various experiences, testing myself in the situations which fortune offered me'. That is how he put it in *Discourse on Method*, his philosophical and autobiographical treatise published in 1637 that contained his best-known line: *Je pense, donc je suis* ('I think, therefore I am'). Though many people may not realize it, mathematics today has been so shaped by Descartes that without his work it would be unrecognizable.

The story goes that Descartes was lying on his back, watching a fly on the tiled ceiling above his bed, while pondering a tricky little geometry problem that goes back to the ancient Greeks. The problem is to find a set of points that satisfy some simple constraints: given two lines, L_1 and L_2, two angles, $\theta 1$ and $\theta 2$, and a ratio, R, find all the points, P, so that $\frac{d_1}{d_2} = R$, with d_1 and d_2 connecting lines as drawn below.

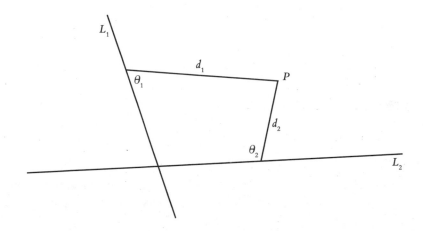

This diagram forced Descartes to start thinking about the best way to express the problem. Sure, he could draw it, but was there another way? As he mulled it over, the fly took off from one tile and landed on another. Moments later, it did the same again. Descartes saw that he could completely describe the whereabouts of the fly by thinking of the tiled ceiling as a grid. If he fixed the starting tile as zero, then he could say the fly had moved *a* tiles in the horizontal direction and *b* tiles in the vertical.

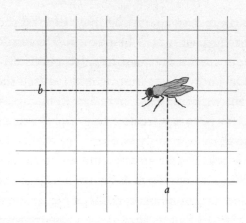

It's worth saying that this story is probably apocryphal, but nonetheless it shows how something as complicated as the movements of a fly can be translated into mathematics. Descartes, one way or another, realized that by thinking of space as a grid he could describe geometric positions and shapes algebraically.

These 'Cartesian coordinates' were a huge breakthrough in mathematics. The system was a bridge between algebra and geometry, allowing problems in one realm to be expressed in the other, doubling the tools available to try to crack them. Take circles, for example.

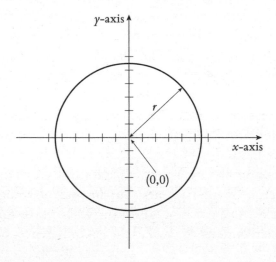

A circle is represented by the equation $x^2 + y^2 = r^2$, where r is its radius.

Mathematicians could draw them, but Descartes' system would allow them to describe them using algebra too. Suddenly, they could use algebraic techniques to answer questions about geometry.

Eventually, Descartes' innovation would become such a standard part of mathematics that it's hard to imagine the subject without it, but at the time few people were able to grasp it. French intellectual Voltaire said that there were only two men other than Descartes who understood the work: Frans van Schooten in Holland and Pierre de Fermat in France. He may have been right about the *men* of the world, but Elisabeth of Bohemia certainly understood it too.

By this point, in 1642, Descartes was in his mid-forties and had moved to The Hague, where Elisabeth was based. The two met at the Exile Court and began corresponding. One letter shows that they were engaged in robust discussions about philosophy. Elisabeth wished to know the reason why Descartes separated the human 'mind' from the 'body'. She wrote to him to ask, 'tell me please how the soul of a human being (it being only a thinking substance) can determine the bodily spirits and so bring about voluntary actions'.[1] Descartes did not have a clear answer. So they corresponded and discussed further on ways of thinking about the mind and body.

They also spoke about mathematics, with Descartes sending her a particularly difficult problem to solve, known as Apollonius's problem, which he believed would reveal the deficits of Jan Stampioen as a tutor. The two were in a quarrel because Stampioen had published a book on algebra and Descartes felt that it trod on his turf. Descartes said that he felt 'rather badly' about sending such a problem to Elisabeth, because he could not 'see how even an angel . . . could possibly solve it without some miracle'.[2] Descartes' condescension was entirely misplaced. Not only did Elisabeth solve it, but she used two different techniques to do so.

Apollonius's problem was not unlike the one that Descartes had been considering involving lines, but instead of straight ones it involved circles. The task was, given three circles, to find a fourth circle whose circumference touches the circumferences of the other three circles.

Elisabeth's first solution used a straightedge and a compass. These

One solution to Apollonius's problem. Three circles are black; a fourth circle, touching the three circles, is drawn in a solid line.

were the standard instruments for mathematicians going back thousands of years and, using several clever tricks, she was able to draw a circle that touched the other three, as shown above. However, the method was somewhat random. Surely there was a better one?

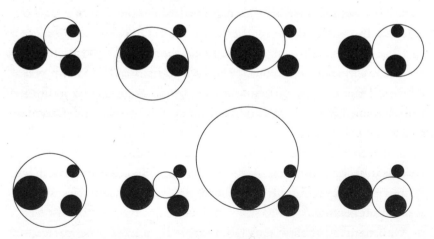

There are multiple solutions to Apollonius's problem.

Elisabeth turned to a different technique that was fresh off the mathematical press: Descartes' very own coordinate system. Recall from above that a circle of radius r has the equation $x^2 + y^2 = r^2$. To tackle Apollonius's problem, Elisabeth worked out that the centre (x,y) of the new circle and its radius r are related to the centres and radii of the three given circles by

$$(x-x_1)^2 + (y-y_1)^2 = (r+r_1)^2$$
$$(x-x_2)^2 + (y-y_2)^2 = (r+r_2)^2$$
$$(x-x_3)^2 + (y-y_2)^2 = (r+r_3)^2$$

where r_1, r_2 and r_3 denote the radii of the three given circles with their corresponding centres having coordinates (x_1, y_1), (x_2, y_2) and (x_3, y_3) respectively, and r, x and y similarly describe the fourth circle.

Each of the + symbols on the right-hand side can in fact be either a plus or a minus. Elisabeth and Descartes didn't deal with the latter cases, but when you take into account those different possibilities, it gives eight lots of three equations, all of which can be solved using algebra to determine the values of r, x and y. In other words, the geometric problem became an algebraic one, revealing not just one solution but all of them.

Clearly, Elisabeth was a talented mathematician. Descartes continued his work and published *Principles of Philosophy* in 1644, which discussed the laws of physics, and dedicated it to her: 'You are the only person I have so far found who has completely understood all my previously published works.'

Elisabeth continued to be fascinated by mathematics and science throughout her life, regularly conversing and debating with the boundary-pushing intellects of the day. Contemporaries such as Dutch intellectual Anna Maria van Schurman, French novelist Marie de Gournay and Irish scientist Katherine Jones (also known as Lady Ranelagh) all became part of her burgeoning intellectual network. Towards the end of her life, Elisabeth moved to an abbey in Germany and conversed with her friends mostly via letters. And so, when she died, she was surrounded by many of her scholarly friends.

8.   The (First) Calculus Pioneers

It was the spectacle of the century. The year was 1715 and a dispute, decades in the making, was finally coming to a head. What started out as a battle about mathematics had engulfed both politics and religion. Each side knew that legacies were at stake and that the outcome would resonate down the generations.

At the centre of the controversy were two mathematicians: Isaac Newton and Gottfried Wilhelm Leibniz. History has chalked up these mathematicians as two of the best. They were prolific and made long-lasting contributions that have, it is no exaggeration to say, changed the world. But at the time of the dispute their places in history were not yet secure. For mathematicians and scientists then, 'priority' over ideas and theorems was everything. Now, it is largely accepted that whoever publishes first in an academic journal has priority, but in the early eighteenth century journals were still in their infancy, so there was not yet an accepted way of settling who got there first. This particular dispute hinged on who had invented calculus.

The implications of calculus were still being worked out in the eighteenth century, but it is probably the most consequential mathematical toolbox we have today. With it, we can study how things change over time. If that sounds vague and non-specific, that's because it is. And that is its strength. There are so many situations in which the way things change is the most important thing to study from a scientific point of view, and this makes calculus extraordinarily powerful. Everything from rocket engines to the way blood flows through your veins can be studied using its techniques. Without them, our understanding of the universe would be a tiny fraction of what it is today.

And so the battle for calculus blew up. On one side was Newton, the darling of English mathematics. On the other was Leibniz, a German polymath who worked as a lawyer and diplomat and had strong royal connections to the House of Hanover. The feud was a matter of national pride — a war that neither side could contemplate losing. Although one of the two would officially be declared victorious, the truth is that both Newton and Leibniz may have been pipped to the post, not just by a matter of years but by a matter of centuries.

The potential post-pipper was a fourteenth-century mathematician called Mādhava. He led an incredible school in Kerala in the south of India that became a melting pot for mathematicians. We are still uncovering exactly how much Mādhava and his school in India knew, but what we know so far is enough to reconfigure our understanding of one of the most consequential breakthroughs in the history of mathematics.

A school in Kerala, India

CALECHVT CELEBERRI-
MVM INDIÆ EMPORIVM.

The view of Calicut, one of the coastal cities in Kerala, in 1572.
From Georg Braun and Frans Hogenberg's atlas *Cities of the World*.

Kerala is a state blessed with fertile land. It lies on the south-west coast of India, facing the Arabian Sea, which it has long traversed for trade with other parts of the world. It is sometimes called the Spice Garden of India, having exported spices to other regions for at least five thousand years. In the fourteenth century, Kerala was made up mostly of small farming communities along the coast. If the growing was good, life was good, but the area was prone to erratic monsoons that could wreak havoc on the crops. Preparing for extreme weather was made more difficult by the fact that the monsoons seemed to strike at random. The residents of Kerala needed an accurate calendar to help them predict seasonal and climate changes. And for this, they turned to mathematics.

Specifically, they turned to Sanskrit mathematical textbooks that had been brought to the region from elsewhere in India. Though many scholars may have been content with writing commentaries and reinterpretations of these works, the intellectual climate in Kerala was more adventurous. New ideas were incorporated into the collective understanding, thanks in part to the region being distant from the politically tumultuous north. Scholars made great efforts to present their findings in locally spoken languages such as Malayalam. Where, in other parts of India, mathematical and astronomical knowledge was reserved for the Sanskrit-savvy upper classes, in Kerala it became more accessible, thanks to a school in Cochin, one of the four main principalities of Kerala, that rose to prominence between the fourteenth and sixteenth centuries.

As far as we can tell, the Kerala school of astronomy and mathematics had no central building, although there was a small-scale observatory that was used to collect astronomical data. Yet, over many years, teachers advanced mathematical knowledge and ideas there, which they then passed on to their students. These students then became teachers themselves. Each generation was responsible for passing on knowledge to the next to keep it alive – a tradition known as *guru–shishya*, literally 'teacher–disciple'.

At the school, teachers disseminated much of the knowledge in the form of verses and phrases, designed to be easy to memorize. However, even if remembering the passages was easy, interpreting them was not. Here is one chant that people used to find the circumference of a circle:

> Multiply the diameter (of the circle) by 4 and divide by 1. Then apply to this separately with negative and positive signs alternately the product of the diameter and 4 divided by the odd numbers 3, 5, and so on . . . The result is the accurate circumference; it is extremely accurate if the division is carried out many times.[1]

The original is in Sanskrit and is much more rhythmic. But even in translation it is easy to see just how sophisticated the resulting mathematics was. In today's notation, we would write the equation the chant describes as follows, where the circumference is written as C and the diameter as d.

$$C = \frac{4d}{1} - \frac{4d}{3} + \frac{4d}{5} - \frac{4d}{7} + \frac{4d}{9} \cdots$$

It's quite extraordinary to see how the verse results in an equation with infinitely many terms, or an infinite series. The more terms you compute, the closer the series gets to the exact value of the circumference. The verse was also a step to obtaining the value of pi, which can easily be seen with a little rearranging:

$$\frac{C}{4d} = 1 - \frac{1}{3} + \frac{1}{5} - \frac{1}{7} + \frac{1}{9} \cdots$$

And then pi is by definition the circumference divided by its diameter.

$$\frac{\pi}{4} = 1 - \frac{1}{3} + \frac{1}{5} - \frac{1}{7} + \frac{1}{9} \cdots$$

Using these same methods, the Kerala mathematicians became proficient in many types of mathematics, including geometry and trigonometry, which they used to help them determine planetary positions and when eclipses would occur.

Most of the mathematicians at Kerala came from the same high-status caste, the Brahmin, which included priests and teachers. Most were younger brothers. Brahmins tended to be the largest landowners and in the group the mathematicians came from, Nambudiri Brahmin, and it was the eldest sons who generally inherited the wealth. Once married, they would be expected to look after family property and community affairs and provide support for their siblings. Younger brothers were not given the same social status or power so they often saw becoming a scholar as a way to establish themselves. Sisters, however, rarely took the same path. In earlier parts of Indian history women of the upper classes had enjoyed similar levels of education to men and there were female intellectuals. However, by the fourteenth century, gender roles had become entrenched, with women for the most part being consigned to homemaking duties.

The Kerala school provided a basic education to those who could afford it and attracted talented mathematicians from across India. The fees they charged funded the school and supported the resident scholars. Most people would be taught by teachers who were non-specialists, but a few of the most talented would become members of the school, meaning they would learn directly from the Kerala mathematicians and join the lineage of *guru–shishya*.

People at the school used palm leaves to record their work, as paper was not readily available. Calendars showing the dates of harvest, local festivals and the new year were often made this way and distributed to Keralan residents. Astrology, which was widely practised across India, provided auspicious times for specific rituals and observances. And that relied on accurate calendars too. But these calendars were not made to last for centuries, so unfortunately many of them have long since perished, along with others from the Kerala school. Only fragments have survived.

We don't know exactly how the school started. One legend has it that it began a thousand years before its rise to prominence in the fourteenth century. The earliest records we have are of a fourth-century figure called Vararuci who came up with rhymes related to the cycle of the moon. Such *chandravākyas* (literally, 'moon sentences') were mnemonics to help determine the position of the moon at a given time of the year. But a millennium would pass before, with the rise of an influential mathematician named Mādhava, the school really started to make waves.

Mādhava was born around 1340 in the village of Sangamagrama in central Kerala. By this time, mathematics in India was already well developed and understood, partly due to the legendary work of true-zero-loving Brahmagupta. Mādhava was Brahmin and spent all his life on his family's estate. His social status and wealth gave him plenty of time to pursue his passion for astronomy and mathematics. He made observations and calculations of the exact position of the moon and started to pass down new knowledge on mathematical calculations to his pupils.

Much of what we know about him is thanks to *guru–shishya*; his works were recorded in book form only later by, for example, his student Paramesvara of Vatasseri, who compiled a textbook in the fifteenth century from teachings passed down through the school. It became a reference guide for efficient agricultural farming such as cultivating paddy fields and dry land. Paramesvara's treatise also included the basic rules of arithmetic, such as how additions and multiplications interact, as well as the eclipses of the sun and the moon and a calculation for the longitude of Kerala.

Two further books, *A Compilation of the Astronomical System* and *Rationales in Mathematical Astronomy*, by mathematicians Nilakantha and Jyesthadeva respectively, gathered many important results that had been passed down from teacher to student. *Rationales in Mathematical Astronomy* is particularly unusual in that it contains proofs of the book's theorems. The astronomical works also show similarities to the work of Danish astronomer Tycho Brahe, discussed in Chapter 9, who lived a century later.

So what of the claims to calculus? For that let us return to the infinite series for calculating $\frac{\pi}{4}$ outlined above.

$$\frac{C}{4d} = \frac{\pi}{4} = 1 - \frac{1}{3} + \frac{1}{5} - \frac{1}{7} + \frac{1}{9} \cdots$$

This is now known as a Mādhava series and is just one of many that the Kerala school knew and understood, suggesting that they hadn't just hit upon it randomly but had worked out some underlying theory. In fact, the proofs contained in *Rationales in Mathematical Astronomy* confirm this. And what was needed to deduce these formulas? Calculus.

Or certainly the basics of calculus. The derivations utilize intrinsic tools in calculus, such as the summations, rates of change and finding the limit – tools that Newton and Leibniz would hit upon centuries later.

What is calculus?

Many people have a fear of calculus. Strange formulas with little obvious meaning are enough to make anyone fret or put you off mathematics completely. This is both understandable and a travesty. Though the details of calculus can be a bit awkward, conceptually it is not only easy to grasp, it is extremely elegant and beautiful. There is something about seeing how neatly calculus comes together that is simply magical.

At its core, calculus is really just a consequence of the idea that big problems are best split into smaller ones. To see this in action, let's begin with a cake. Our cake is a perfect cylinder (because it was baked by Kate and not Timothy) with a radius *r*.

Beyond how it tastes (great!), an obvious thing to ask is: how much cake is in our cake? In other words, what's the volume of our cake? Inspired by the idea that big problems are best split up into smaller ones, let's try cutting it up. If we then put it back together in a different way, we can see that from above it looks like a parallelogram, but with a curvy top and bottom.

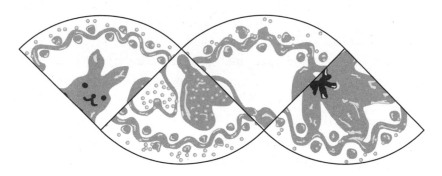

Well, a bit like a parallelogram. But if we cut the cake into smaller pieces, it does then start to look more like one.

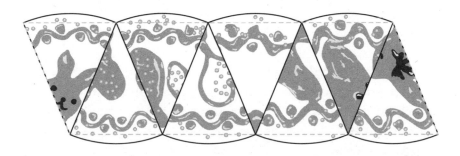

Calculus's leap was to keep cutting that cake, making the pieces smaller and smaller. With each decrease in the size of the slices, the curves along the bottom and the top become less pronounced, meaning that if we could cut the cake infinitely many times, we would end up with a fully bona fide parallelogram.

Of course, in real life we can't cut a cake into infinitely many slices, but mathematically we can, and so we can conclude that the area of the top of the cake is equal to the area of this parallelogram.* And,

---

* Dealing with infinity is a tricky business. It works out nicely here, but there are plenty of examples where it doesn't. So rather than just 'imagining' what happens, there is a rigorous theory underpinning it all to make sure things don't go wrong. The subject is called 'mathematical analysis' if you are interested in reading more.

luckily for us, the area of a parallelogram is easy to calculate: it's just the width multiplied by the height.

The height of this particular parallelogram is the radius of the cake minus a little bit – the difference between a straight vertical line reaching the dotted line or the curve. Let's call this difference dr.

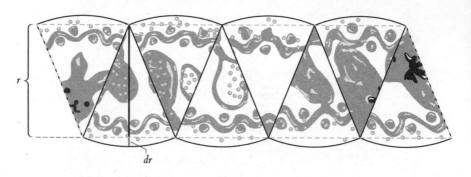

Measuring dr would be a bit of a pain, but luckily we don't need to worry about it too much. It's clearly a small number, but the smaller our slices, the smaller dr becomes. And as we approach infinity, dr approaches zero. This means that the height of the parallelogram can be treated as equal to r.

And the width of the parallelogram is half the circumference of the cake. If you look at just the top, for example, you can see it is made up of the edges of half of the slices. The circumference of a circle is given by $2\pi r$, so our width, as half of this, is πr. Therefore, with width times height, our parallelogram has an area of πr^2.

Of course, this is the formula for the area of the circle, just as we would expect. But the way that it pops out of this process is nothing short of remarkable. Rather than regurgitating it, we have derived the area of a circle using the principles of calculus – and confection.

So all that's left is to multiply by the height of the cake, h, to get the formula for the volume of our cake, $\pi r^2 h$. Now, not only do we know some calculus, we also know how much cake is in our cake.

But we can take this further. As the pieces we used to approximate the size of the cake got smaller and smaller, we got closer and closer to the exact size of the cake – that's the brilliance of calculus. The idea of imagining what would happen as we approached infinity is known by mathematicians as finding the limit, and the outcome here was a technique called integration that is extremely versatile for finding the areas of graphs.

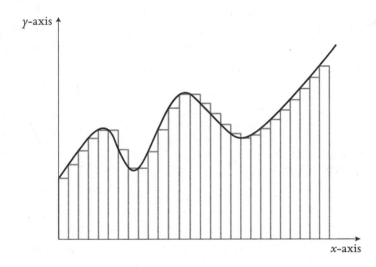

Using the same approach, the area under a graph like the one above can be approximated using simple shapes like rectangles – the more rectangles you have, the closer you get to the exact area under the graph. Normally, mathematicians write the widths of the little rectangles as dx and, if we call the curve being shown $f(x)$, then the integral is written as $\int f(x)dx$.

These symbols might look familiar to you, or they may not, but they are not especially important to a conceptual understanding of this process. For that, all you need to understand is that when calculating the area of a graph there is a calculus technique called integration that uses infinitely many approximations – and that calculating the area under a graph is often very useful.

Calculus has one more fundamental concept we have not yet touched on. There are limits, integration and, finally, differentiation. This last one is a sort of reverse process to integration (this relationship is known as the fundamental theorem of calculus) and uses limits to work out the gradient of a curve.

To see this in action, let's take a trip on a Japanese bullet train. These travel at speeds of up to 320 km/h. The faster they can reach their destination, the happier their customers. However, the customers can be picky. If the train gets up to full speed too fast, everyone (and the drinks cart) is flung to the back, leaving them squashed (and soaked). So to prevent this from happening the train company ought to make sure the train's speed doesn't change too quickly. But how can they accurately measure this?

Here's a graph of one approach for getting the train moving quickly. Accelerate fast at the beginning and then slowly head towards top speed.

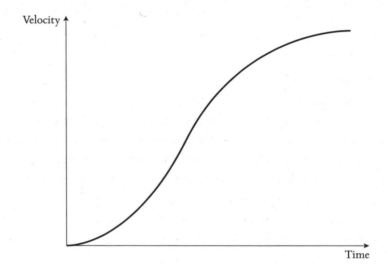

To measure the acceleration of the train at any point, the train company needs to know how quickly the speed of the train is changing, which is captured by the slope of the graph. But measuring the slope of curly graphs is tricky, so one way to approximate this is to draw a straight line between two points and measure the slope of that instead.

Now we've got an approximation, but the smaller the gap between the two points we choose, the closer we get to the exact slope at the point we're interested in (and therefore the acceleration at that time). If we do this to infinity, that is, we find the limit, we get the exact

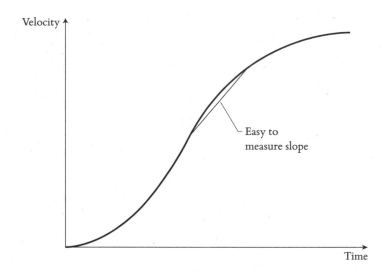

value of this rate of change, often written as $\dfrac{dv}{dt}$ to denote the change in velocity over the change in time.

These tools of differentiation, integration and finding the limit are incredibly versatile. They are used for everything from calculating how a business can best make a profit and measuring whether medical drugs are having an effect to tracking how the planets move.

Newton tries to get a book deal

And this brings us back to Newton and Leibniz. In 1665, Newton was on the verge of a breakthrough. He was still in his early twenties and had just graduated from Trinity College, Cambridge, where he had studied a mixture of subjects, among them philosophy, mathematics and astronomy. He was ready to dedicate himself to research in Cambridge, but the plague had struck England, with the country in the midst of the worst outbreak since the Black Death of 1348. London alone would lose nearly a quarter of its population. And so the university was closed.

Newton went back to his hometown of Woolsthorpe in Lincolnshire in the East Midlands, where he had little in the way of family to rely on. His father had died three months before he was born and his mother had abandoned him when he was just three years old, on the insistence of her new husband. She had left him to live with his grandmother, which he continued to hold against his mother. He believed his grandmother was rich enough to support him but chose not to. Instead, before he won a scholarship, he had to be a sizar – a person who earned their keep as a servant to wealthier students. So, Newton was all by himself when he returned to Lincolnshire and stayed at the home of Mr Clark, the town pharmacist.

Mathematically speaking, he was in the prime of his life. He later wrote that he was more focused during this time in Lincolnshire than at any other time afterwards.

He began by studying the work of seventeenth-century European mathematical giants such as Frans van Schooten, René Descartes and Pierre de Fermat, among others. From their books, he learned the foundations of Renaissance mathematics and Descartes' coordinate system. He also learned about ways to calculate tangents. These would be crucial to calculus because they represented the rate of change of a curve. From his notes, it appears Newton spent much of his time thinking in the abstract, but he also probably had in the back of his mind the problems in physics that we know interested him, for example the way an apple falls from a tree* or a planet orbits the sun. These both concern the rate of change, and no mathematics at the time could adequately deal with these situations.

In Europe, mathematicians before Newton had come up with ways to determine tangents, but the methods they came up with weren't particularly robust, working in some circumstances but not others. By understanding when these methods worked and when they didn't, he developed his own method, which was completely general. It didn't just work for some curves, like his predecessors' methods, but for any curve. Newton already knew at this point that tangents and areas under curves are opposites. So by making progress on tangents, he was also making progress on determining the area of a curve, another crucial part of calculus.

---

* As far as we know, one never actually landed on his head.

He gathered these ideas into two books. But like many aspiring authors, he initially didn't have much luck finding a publisher. The Great Fire of London in 1666 had devastated publishers so they could no longer afford to issue mathematics books, which were foolishly viewed as 'slow selling'.* However, while Newton's manuscripts were lying dormant and unpublished, someone else was busy making discoveries about tangents and the areas under a graph of their own.

Royal Society portraits of Newton (*left*) and Leibniz (*right*).

Leibniz makes a maths machine

Gottfried Wilhelm Leibniz was a precocious child. In 1661, aged fourteen, he was already studying philosophy at the University of Leipzig. Although this seems very young, it was not that unusual at the time. What is more surprising, given his place in history, however, is that his university was weak on mathematics. His course involved plenty of material on rhetoric, Latin, Greek and Hebrew

* Thankfully, things have changed!

but was lacking when it came to the good stuff. He would have to pick up mathematics on his own.

Leibniz completed his two-year degree, went on to study law and received a doctorate in the subject in 1667. After graduation he started his legal career, living in various places in Germany, including Mainz, where he eventually turned to diplomacy and became a secretary to the politician Johann Christian von Boineburg. He threw himself headlong into the world of diplomacy, culminating in an audacious plan towards the end of 1671.

Hearing about the intention of the ambitious Louis XIV of France to invade the Netherlands, he decided to try to convince Louis to invade Egypt instead. He hoped that this would help protect Germany from any potential war or invasion. Leibniz made his way to Paris to make contact with the French government as a representative of his local governor. Due to the Franco-Dutch War in 1672, it took some time before he had an opportunity to approach the government there in Paris, so in the meantime he started studying mathematics and physics with Dutch polymath Christiaan Huygens. Although he had little luck with his Egypt diversion plan (Louis XIV invaded the Dutch Republic), Leibniz had now become enchanted by mathematics too.

He made a trip to London on diplomatic business in 1673, and used the opportunity to contact mathematicians at the Royal Society to show off a recent invention he had made that helps to explain Leibniz's mathematical motivations. The invention was the first calculator that could perform the four arithmetic operations: addition, subtraction, multiplication and division. Called the stepped reckoner, it was a contraption based on a series of gears and dials that, once set, could perform arithmetic. Although the design was good, building it was largely beyond the capabilities of the time, so the machines didn't work reliably. However, the central mechanism, known as a Leibniz wheel, was used for calculating machines for hundreds of years. The Royal Society was impressed, electing him a fellow in 1673.

Leibniz turned to geometry. He focused on improving methods for calculating the area under a curve. Just like Newton, Leibniz found that existing mathematics worked in some cases but not in

Stepped reckoner.

others, and he managed to develop a more general theory of calculus. This approach was the opposite to Newton's but equally effective. By making progress on calculating the area under a curve, Leibniz made progress on finding tangents.

His motivation was less about physics and more about mechanizing the rules of logic. Just as with his stepped reckoner, Leibniz aspired to make machines that could perform all manner of calculations. This led him to describe a sort of arithmetic for his methods of finding the area under a curve and tangents. He had hit upon various rules for differentiation and integration and in the process he simplified the notation in a way that has not been improved upon since. Over just a couple of weeks, he coined the symbols and conventions we still use today, hundreds of years later. That large S-shaped symbol that denotes integration? Leibniz's doing. Leibniz and Newton had hit upon essentially the same mathematical idea, just presented in different ways.

Olde English snark

Newton and Leibniz didn't exactly have an amicable relationship. By the early eighteenth century, they had exchanged several letters, but they mostly consisted of Leibniz trying to impress Newton and Newton responding with indifference, mockery and outright hostility.

In one instance, Leibniz sent Newton some of his results on infinite series – like those known by the Kerala school. Leibniz was hoping to impress but, instead, what he got in return can only be described as olde English snark: 'three methods of arriving at series of that kind had already become known to me, so that I could scarcely expect a new one to be communicated'.[2] In other words, not only did Newton already know the mathematics that Leibniz had sent him, he knew of three other ways to reach the same conclusion. Leibniz was late to the party.

Newton's snark continued. Rather than finishing the letter by sending Leibniz some mathematics on the underlying theory that would help him out, he wrote: 'The foundation of these operations is evident enough, in fact; but because I cannot proceed with the explanation of it now, I have preferred to conceal it thus: "6accdae-13eff7i3l9n4o4qrr4s8t12vx." '[3] This was a taunt. The strange string of letters and numbers was a coded message for the fundamental theorem of calculus in Latin.

At this point, it appears that neither Newton nor Leibniz knew the extent to which the other had developed his own version of calculus. This changed when Leibniz published a paper outlining his ideas in 1686 in the scientific journal *Acta Eruditorum*. Though it's not clear if and when Newton read this, other mathematicians certainly did, making public declarations of foul play, proclaiming that Leibniz had stolen Newton's ideas. The marquis de l'Hôpital said that, although he preferred Leibniz's notations, Newton should get full credit for the work. Another mathematician, Nicolas Fatio de Duillier, suggested that Leibniz had committed plagiarism, declaring that 'Newton was the first and by many years the most senior inventor of calculus.'

Leibniz did not stay quiet. By this point he was the first president of the Berlin Academy of Science and a famous intellectual. Receiving a printed copy of one of Newton's newly published books, *Opticks*, Leibniz published a review anonymously in which he wrote that Newton had made 'elegant use' of Leibniz's ideas. His friend the

Swiss mathematician Johann Bernoulli came to Leibniz's aid and attacked Newton, saying that he had not used the techniques of calculus in his earlier book *Principia* and so therefore could not have been first to them. Then John Keill, a Royal Society mathematician, entered the fray. He defended Newton's work and said that 'the same arithmetic . . . was afterwards published by Mr Leibniz in the *Acta Eruditorum* having changed the name and the symbolism'.[4] Because Leibniz had become a member of the Royal Society and Keill was a fellow there, Leibniz asked the Society for a retraction.

A committee was formed for members to decide once and for all who had the right to call calculus their own. Unfortunately for Leibniz, however, Newton had an inherent advantage: he was the president of the Royal Society. In March 1712 Newton gave the committee documents he believed established that he had got there first. He then drafted the report on behalf of the committee for a wider circulation. It, unsurprisingly, concluded: 'Mr Newton was the first inventor.'[5]

However, this was not the end of the dispute. The matter turned from being one about mathematics into one of politics when Caroline, Princess of Wales, entered the fray.

Originally from Germany, Caroline was orphaned as a teenager and so grew up in Hanover, under the care of Sophia Charlotte, the Queen of Prussia. Leibniz had been Sophia Charlotte's personal tutor and she passed on his teachings to Caroline. Caroline and Leibniz later became friends themselves. When Sophia Charlotte died, Caroline became a patron of Leibniz in turn. Their intellectual friendship grew strong and continued until Caroline and her husband, George, moved to London following the death of Queen Anne in 1714.

Now Caroline was in London, prominent intellectuals sought to distance her from Leibniz and urged her to support Newton instead. The influential philosopher and cleric Samuel Clarke met with Caroline in person to explain the work of British mathematicians to her and brought Newton along too.[6]

The feud between Newton and Leibniz turned nastier, fuelled by their opposing ideas about religion – Leibniz believed in the

Caroline, Princess of Wales. Painted by Godfrey Kneller in 1716.

reunification of the Protestant churches; Newton did not. Leibniz wrote to Caroline apportioning the blame for the religious decay he saw in England to Newton and his followers; she showed the letter to people in London, among them Samuel Clarke, and it was taken as a major insult.

Caroline didn't take sides. She often argued with Clarke and even Newton, while continuing to exchange opinions with Leibniz via letters. Her wish was not to claim a victory for Leibniz; she worked as an arbiter and a moderator, trying to find the common ground and make peace between them. She wrote in April 1716, 'I am in despair that persons of such great learning as you and Newton are not reconciled. The public would profit immensely if this could be brought about.'[7] Unfortunately, a reconciliation never was brought about. Both Newton and Leibniz continued to maintain until their deaths that they had each got there first.

The verdict

It is now widely accepted that Leibniz and Newton worked on calculus independently of each other, building on the work of others who came before them. History has focused on their stories, presenting this as an epic fight between two geniuses, with shouts of plagiarism sprayed indiscriminately. It is wrong to claim that the origins of calculus lie with either Leibniz or Newton, as one thing is certain: neither of them got there first. But it is also wrong to give the Kerala school all the credit.

Calculus is a wide and varied toolbox and as such it has wide and varied origins. Newton himself famously said, 'If I have seen further it is by standing on the shoulders of giants.'[8] The giants were the mathematicians and teachers that came before him and passed on their work to subsequent generations. This is always the case with any mathematical development. Progress in the subject is a long, winding walk towards truth and we should be mindful to take seriously the contributions of those who took steps along the way, not just those who took the final ones.

In the early nineteenth century, when English civil servant Charles Whish first brought the work of the Kerala school to light in the West, he was met with scepticism. He had come across it while working in Madras. One of his hobbies was to collect palm-leaf manuscripts* and, as a linguist, he could read both Sanskrit and Malayalam. In a paper in 1834, he compared the collective knowledge of Indian mathematics up to that point, from the Kerala school and elsewhere, with European mathematics of the same period. As far as we know, he was the first to do this and, what's more, he concluded that Indian mathematics was more advanced. This greatly challenged the Eurocentric view of scientific development and anti-Indian bias. However, at the height of European imperialism, few people were willing to look at the evidence.

---

* Of these, 195 are currently in the possession of the Royal Asiatic Society of Great Britain and Ireland.

More recently, mathematician George Gheverghese Joseph, who was born in Kerala, has sounded the alarm. He has made the point that, too often, histories of mathematics focus on Europe and he highlights the dangers of continuing the view that European mathematics was the most advanced in the world. In *A Passage to Infinity* (2009) he argues that there was a pathway for knowledge from India to the West. So, rather than developing calculus in Europe in an isolated and independent way, Leibniz and Newton could have been influenced by the school in Kerala. This claim is not fully underpinned by existing evidence and is still under investigation. However, his idea that mathematics has many origins has paved the way for historians to think further about the non-Western roots of mathematics.

There is still work to be done on understanding exactly who knew what and when. But revisions to our view of the history of mathematics should not come as a surprise. Many Western scholars have long held the blinkered Eurocentric view that people outside the West had no interest in science, mathematics and the world. And nor should it come as an unwelcome challenge. The idea that a school in India could have passed the baton to Newton and Leibniz is an exciting possibility. It is also one that would very much fit with the beautifully chaotic way in which mathematics progresses. Though mathematics is often presented as consisting of neat, logical sequences of ideas, proofs and theorems, its history is rarely so straightforward.

9.   Newtonianism for Ladies

At the end of the sixteenth century, Mars was causing a bit of a problem. The Danish astronomer siblings Sophia and Tycho Brahe had made thousands of incredibly detailed measurements of celestial bodies from the Swedish island of Hven, which was then under Danish rule. Tycho had received significant backing from King Frederick II of Denmark and Norway and had built arguably the world's finest observatory there. He was attempting to confirm which model of the solar system was correct: Ptolemy's, which put the Earth at the centre; or Copernicus's, which had the sun at the centre. Sophia was one of Tycho's most trusted colleagues.

Neither view seemed to be able to properly account for Mars.

Tycho Brahe and Sophia Brahe. The pair had eight other siblings and were around ten years apart in age. Sophia was widowed with a son and studied horticulture, chemistry and astronomy. She was taught by Tycho but also bought astronomy books written in Latin, which she had translated into German.

Predictions of where the planet should be throughout the year would later be found to be off by several degrees. Precision was incredibly important to the Brahes – it couldn't have been human error, could it? Finally, Tycho gave up and passed the Mars problem to one of his assistants, a young Johannes Kepler.

Kepler was diligent and precise, but he was also more willing to challenge fundamental assumptions than his mentor. Both Ptolemy and Copernicus had insisted that the orbits of the celestial bodies were formed from perfect circles, but Kepler wondered if this was really true. He re-examined the observations and, much to his surprise, switching to elliptical orbits – effectively, flattened circles – reduced the error by a factor of ten. The idea of elliptical orbits had first been proposed by court astronomer al-Ṣābiʾ Thābit ibn Qurrah al-Ḥarrānī in the ninth century at the House of Wisdom, though it had not yet been widely accepted.

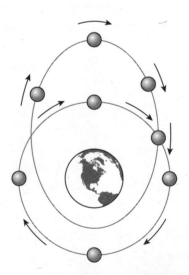

Circular and elliptical orbits.

Up to this point, the prevailing view was that circles were in some way a divine shape. Now, elliptical orbits were matching Kepler's observations. To find out why, the world needed calculus.

Part of Newton's original motivation for developing his form of calculus was to better understand the universe. Before him, many people believed that it was perfectly reasonable to describe different forms of motion differently. Why should the laws that describe the way the planets move be the same as those for the way that an apple falls from a tree? But Newton believed otherwise. 'To the same natural effects, we must, as far as possible, assign the same causes,' he wrote in his magnum opus, *Principia*.[1] In other words, motion is motion, no matter what.

This principle led him to come up with the laws of motion that are at the heart of Newtonian mechanics. These describe the basic principles that Newton believed all motion follows. From those, he was able to derive the Law of Universal Gravitation – essentially an equation that describes how objects act under the influence of gravity. This equation showed categorically that Kepler was right: the planets do follow elliptical orbits because they satisfy the laws of motion.

Newtonian mechanics was an extremely powerful mathematical theory that would drive physics for hundreds of years. But overturning the consensus rarely happens swiftly. Convincing others of the usefulness of Newtonianism would take decades – and an unusual host of characters.

Newtonianism 101

Before we get on to the journey that Newtonianism took, let's take a moment for a quick reminder of what it involves. Newton's laws of motion, as paraphrased by NASA,[2] are as follows:

Law 1 An object at rest remains at rest, and an object in motion remains in motion at constant speed and in a straight line unless acted on by an unbalanced force.

Law 2 The acceleration of an object depends on the mass of the object and the amount of force applied.

Law 3 Whenever one object exerts a force on another object, the second object exerts an equal and opposite on the first.

Newton's first law essentially says that a force is always needed to move a stationary object or to change the way a moving object is moving. This is straightforward enough, but it may leave you wondering why it shouldn't imply that if you roll a ball across the floor it continues rolling for ever. Forces aren't always easily spotted. In this case, a ball rolling on Earth would be slowed down by the forces of friction and air resistance. If you pushed a ball in space, it would keep on going for ever.*

The second law is often written as the famous equation $F = ma$, denoting that the force needed to move an object is equal to its mass times the acceleration you wish to impart. This is where calculus comes in. Acceleration is a measure of the rate of change of velocity, so the equation can also be written

$$F = m\frac{dv}{dt}$$

If you were to draw a graph of velocity against time, $\frac{dv}{dt}$ tells you the tangent at any given point.

The third law is one that you will know well if you've ever walked into a wall. Sure, you apply force to the wall, but it hurts as if the wall applied the force to you.†

Newton applied these laws to see what happens in the case of gravity and celestial bodies, taking another observation that Kepler made and combining it with his second law. Kepler's observation was to do with the relationship between the size of an orbit and the time it takes to come full circle; specifically, the square of the size is proportional to the cube of the time. Using this idea, the second law of motion and some manipulation of equations, Newton derived his Law of Universal Gravitation, written as

$$F = G\frac{m_1 m_2}{r^2}$$

This gives the force between two objects with mass m_1 and m_2 that are separated by a distance r. The letter G is the universal gravitational constant, a fixed value related to the strength of gravity in our universe.

---

* Or until another force, like gravity, affected it.
† We do not recommend testing this. If you have not previously had the experience, you'll just have to trust us.

From this equation and others, Newton could deduce more precisely than ever before how orbits work. As an orbiting object moves away, the force produced by gravity decreases exponentially. Doubling the distance quarters the gravitational force; tripling it drops it by a factor of nine; and so on. This means that planets orbiting the sun get gradually further away before turning back in the other direction when gravity is at its weakest. In very rare cases, if the velocity is exactly right, this can lead to a circular orbit – Venus's orbit, for example, is near circular – but otherwise the orbits are ellipses, with Mars, Tycho and Sophia Brahe's original problem planet, being one of the most striking examples.

Flat-ish Earthers

Newtonian mechanics caught on quickly in England. An influential group in the Anglican Church felt Newton's ideas matched their own doctrines that the universe was governed by 'divine' laws. Fellows of the Royal Society of London supported Newton's work too, and when he became president of the Society in 1703 this enabled his work to spread even further.

But Newton's theory wasn't met with universal acclaim. The Swiss mathematician Johann Bernoulli rejected it completely, believing that the idea of a force that could act across empty space was 'unintelligible'. In France, many people preferred Descartes' view of the solar system – that the Earth, moon, planets and stars were immersed in an invisible fluid called ether. The Cartesians believed that God had placed this ether there at the dawn of time. Followers of Descartes believed that reason was fundamental to knowledge whereas followers of Newton held empiricism and mathematics in greater esteem.

With the fall of the Anglo-French Alliance in 1731, tensions across the Channel heated up, as did the scientific debate. Take Newton's claim that the Earth is flat. Well, not flat flat, but a little flat at the top and the bottom. Working through his ideas on motion and gravity, Newton had come to the view that Earth could not be a perfect sphere. The spinning of it on its axis would cause the equators to experience a stronger outward force than at the poles, meaning Earth would swell there. Newton believed this would make Earth an oblate spheroid rather than a sphere.

Earth has the shape of a flattened sphere known as an oblate spheroid.
The lower drawing shows an exaggerated version of this.

To work out whether Newton or Descartes was right, the French Academy of Sciences sent a mission to the equator and another towards the North Pole. It was the first global multi-city collaboration to prove a scientific idea through experiments.

The expeditions had two methods to help them determine the shape of Earth. First, they measured the speed of a pendulum clock in the different locations. The stronger the gravity, the faster the clock should tick. Second, they checked the stars. By measuring the same stars in each location, the teams could work out whether they were viewing them from angles that would confirm whether Earth was a perfect sphere or not.

But deploying these methods successfully turned out to be far from easy. The North Pole group was led by renowned French mathematician Pierre-Louis Moreau de Maupertuis. He had studied in London for several months and was a Newtonian who was keen to prove the effect of gravity. However, as he was not a trained astronomer he asked Swedish astronomer Anders Celsius to join him and his team. Before setting off, they acquired several astronomical instruments in London, including a custom-made telescope made by the skilled instrument maker George Graham.

Their expedition to Lapland, near the North Pole, presented challenges. They had to take measurements from the frozen surface of the Gulf of Bothnia, the northernmost arm of the Baltic Sea, and the lack of accurate maps caused problem after problem. Their brand-new instruments were cumbersome, so the team decided to

A drawing of the custom-made telescope made by George Graham.
From Pierre-Louis Moreau de Maupertuis's *Degree of the
Meridian between Paris and Amiens*, published in 1740.

stay in Torneå in northern Finland. They built a small observatory, took measurements and lodged in housing provided by the local people. After a harsh winter, they completed the measurements within a year.

Though the expedition was difficult, it was nothing compared to the one their counterparts at the equator undertook. French astronomers Pierre Bouguer, Charles-Marie de La Condamine and Louis Godin left France with their assistants – some of whom were enslaved people – and two Spanish naval officers in 1735. The group sailed to the Caribbean coast, travelled overland across Panama and took another ship to the Pacific Peruvian coast, to arrive in New Granada, then a Spanish territory.

The team was small, but the three scientists had large egos and didn't get along; they split up way before they reached Quito, in today's Ecuador. Team leader Godin took most of the money and equipment, leaving Bouguer and La Condamine to make the journey by themselves. After sailing down the Peruvian coast on a boat, recording a solar equinox and a lunar eclipse, Bouguer and La Condamine took two different paths to Quito. Bouguer headed for the mountains, walking along the Andean volcanoes, and La Condamine went straight through the tropical jungle.

When the group members finally reunited in Quito, they started to take the measurements, but working at high altitude was not easy. The weather was severe and they had to stop and wait for months for a clear sight of distant reference points. Wild bears and poisonous snakes presented difficulties and some of the team brushed with death more than once. Several caught malaria, and one died of it. Another lost his life in a street fight.

The three-year project became a nine-year project, and they eventually called it a day in the spring of 1743. Bouguer and La Condamine travelled back to France separately, using the remaining money from the French Academy of Sciences to fund their journeys and leaving inadequate funds for the other members of the team. They simply abandoned their assistants and helpers. The Spanish naval officers made their way back thanks to funding from

An illustration in *Observations Astronomical and Physical*, a book published in
1748 outlining the results from the expedition to Quito. The globe is
clearly oblate, flattened at the poles. The Spanish naval officers who
accompanied the group, Antonio de Ulloa and Jorge Juan y Santacilia,
published their results from the expedition in 1748 in Madrid
before the French explorers published theirs in France.

the local Spanish authority. Some of the team died in South America, while others spent up to fifteen years in the Amazon trying to
raise money to get home.

When the data from South America and Lapland finally made it
back to the French Academy of the Sciences, it showed that Newton
was right: Earth is an oblate spheroid. This helped tip the balance for
many naysayers, and it showed just how successful Newtonian
mechanics was at making predictions and kickstarted the spread of
his ideas across the continent.

What's French for Newtonianism?

One particularly important proponent of Newtonianism was Émilie du Châtelet. Like many eighteenth-century noblewomen, du Châtelet had received her education from private tutors. Her parents wanted intellectual freedom for their children and had encouraged her to express her opinions on a wide range of topics at home and during the weekly salons they held. Her father was Louis XIV's chief of protocol at the Palace of Versailles, giving the family enough status to live such a life.

Émilie du Châtelet.

In her twenties, she received tuition from Pierre-Louis Moreau de Maupertuis, who would later lead the Lapland expedition and was a supporter of Newtonianism. He taught du Châtelet algebra and

Newton's version of calculus, further boosting her love of academic discussions in the salons.

Mathematicians such as Maupertuis would often gather at the king's library or in Café Gradot in Paris, but du Châtelet wasn't allowed to join them because she was a woman. Unwilling to tolerate such exclusion, on one occasion she rocked up to a meeting dressed as a man and was promptly let in.

It was around this time that a familiar face came back into her life. She had met the prolific writer and philosopher Voltaire* at her father's salon when she was younger but, due to his vocal criticisms of the French government, he had been forced away – twice sentenced to prison and once sent to England. In 1733, he returned from his exile and they met again.

Even though du Châtelet was married, she and Voltaire retired to her husband's estate in north-east France and lived as a couple there from 1735 to 1739. Affairs were fairly common and often tolerated because marriage was considered a formality and a duty for aristocratic families at the time. Voltaire spoke highly of his partner in public, citing her as living proof that women were just as capable as men.

In 1739, when the French Academy of Sciences announced a prize for the best response to the question 'What is fire?' both du Châtelet and Voltaire submitted their responses independently; Voltaire was unaware that du Châtelet had responded until the list of entrants was published.

Neither of them won the prize, but du Châtelet's paper put forward a remarkable insight. It hypothesized that energy should be conserved in a system. She proposed that, if two objects moving at speed smashed into one another, some of the energy would turn into heat, some into sound and some would remain as kinetic energy. If you added it all up, du Châtelet proposed, it would be equal to the amount of energy before the objects collided. This is now known as the law of conservation of energy and is a fundamental rule of the universe. German mathematician Emmy

---

* His birth name was François-Marie Arouet, but he's normally known by this nom de plume.

Noether would find a mathematical basis for it one hundred and fifty years later (see Chapter 12). It was a profound leap, as understanding of the concept of energy was in its infancy at the time.

At the request of René Réaumur, one of its members, the Academy published du Châtelet's paper and Voltaire's paper, along with the winning entries. With this, she became the first woman to have an original scientific paper published.

Title page of *Dissertation on the Nature and Propagation of Fire*.

One of du Châtelet's first complete works was a physics textbook written for her thirteen-year-old son. In *Lessons in Physics*, she compiled the ideas and theories of Descartes, Leibniz and Newton. Originally published in 1740, anonymously, to conceal the fact that it had been written by a woman, it was the first new publication on physics written in French since 1671.

Du Châtelet was fascinated by Isaac Newton's work as outlined in

Principia and embarked on a project to translate it into French. She had to work around raising three children and so it took her around four years to finish. Often, her most productive writing hour was at four or five in the morning.

Unfortunately, she didn't live to see it published. Du Châtelet fell pregnant by poet Jean-François de Saint-Lambert. She gave birth at the age of forty to a daughter but developed a fever some days later. She knew her time was near, so asked for the manuscript to be brought to her bedside and wrote on it '10 September 1749' as its completion date. Soon after, she lost consciousness.

Seven years later, in 1756, the book was published, though only in part. It was widely praised. Her translation expunged jargon and she explained basic terms such as 'orbit' and 'ellipse' in ways that beginners could understand, often using analogies to clarify the meaning.

By the time of publication, there was a renewed interest in France in Newtonianism, following the successful prediction of the return of Halley's comet in 1759. As a result, the first full French version of *Principia* was published in 1759. Du Châtelet's book is considered the standard French translation of the work to this day.

Experiments at home

In Italy, it was Laura Maria Caterina Bassi Veratti who was at the forefront of Newtonian physics. Born in Bologna in 1711, Bassi had a wide-ranging education, from classical languages to natural philosophy. Her father hosted salons, inviting the top scholars of the city to their home. Her tutor, physician Gaetano Tacconi, introduced her to members of the Bolognese scholarly community and she soon became known as a child prodigy. Her reputation was cemented when a cardinal convinced her to publicly debate with four or five professors. The debate was a triumph for her. She successfully defended forty-nine texts on philosophy and physics in front of dignitaries, including Pope Benedict XIV. The sixteen members of the Academy of Sciences of Bologna unanimously agreed to make her a member in 1732.

Italy was generally more liberal for women than other parts of

Europe, allowing them to access university education. Bassi received her doctorate in philosophy and, on joining the University of Bologna, she became the world's first female professor.

At that time in Bologna, some academics remained Cartesian while others were firmly Newtonian. Bassi was in the latter camp. She taught courses on Newtonian physics for twenty-eight years, but one major difficulty was that only men were consistently allowed to give public lectures. She was rarely given permission to lecture – only when a distinguished visitor came to the city or a degree was conferred on someone. She was paid a salary by the university but given no real academic tasks to perform.

However, Bassi found a way around this. She and her husband, fellow professor Giuseppe Veratti, held a 'literary conference' two evenings a week at their home. As a married woman, she was able to invite guests and students and lecture there without consequence. Bassi also built a home laboratory in order to be able to perform experiments, the results of which she presented to her academic colleagues. Many of these reports have not survived, but we know they touched on electricity, gases, mechanics, fluid dynamics and optics.

Bassi kicked off a new movement to study Newton's ideas. Her fame spread and students and visitors came from Greece, Spain, Germany, Poland and France. She presented the results of her home experiments at the Academy of Sciences of Bologna and published papers whenever she could find a publisher. She was the only woman at the time to be publicly announcing original results in experimental physics. She became the special chair in experimental physics at the University of Bologna, with her husband as her assistant. She petitioned to increase her salary to better equip her home laboratory. When the university said yes, she became the highest-paid professor at the university, investing the money into the lab.

Too much imagination

Considering the work of du Châtelet and Bassi, it's strange to think that the next big development in the spread of Newtonian mechanics was a book aimed specifically at women that implied they needed some extra help. Nonetheless, *Newtonianism for Ladies*, originally published in 1737, was a resounding bestseller and drew the attention of both women and men to Newton's *Principia*.

The author was eighteenth-century Italian polymath Count Francesco Algarotti. He believed, typically of the time, that women were simply 'too imaginative' for mathematics. Even if it were true that women have more imagination than men by some measure, that would hardly be a disadvantage. Mathematics is all about imagining things that don't really exist. Even something as simple as a circle – a mathematically perfect one – does not exist in real life. However hard you tried, you would never be able to build one or draw one. Instead, you have to use a little imagination.

In his book, Algorotti tried to make Newton's ideas reader friendly by using fictional conversations between a chevalier and a marchioness, modelled after himself and du Châtelet. The location used as a backdrop was inspired by du Châtelet's countryside home, which Algorotti had once visited. On top of this, Algorotti made Newtonianism a little romantic. 'I cannot help thinking . . . that this proportion in the squares of the distance of places . . . is observed even in love,' says the marchioness at one point. 'Thus after eight days absence, love becomes 64 times less than it was the first day.'[3]

Steamy stuff.

Just as with the law of universal gravitation, the marchioness is saying that love also decreases as an inverse square as the time of separation becomes longer. Absence makes the heart grow colder.

Algorotti's book was not the only book on Newtonianism aimed at a general audience. *The Ladies' Diary*, for example, was a London-based annual publication that featured important calendar-style information such as the lunar cycles and the dates of school terms. It also included puzzles that became increasingly challenging – and some required

In *Newtonianism for Ladies* (1737), six dialogues took place on five
consecutive days. The backdrop looks similar to Cirey, France,
where Algorotti met du Châtelet and Voltaire.

calculus to solve. In addition, Italian mathematician César-François
Cassini de Thury wrote a book in the 1740s in the form of a dialogue
between a man and a woman that introduced the debate about the
shape of the Earth. Newton's ideas were going mainstream.

Coming to America

Newtonianism then set its eyes on America. British, French, Dutch
and Swedish settlers had landed on the East Coast in the early sixteenth and seventeenth centuries. Many Europeans moved to America
to escape the political, economic and religious hardships they faced
in their homeland. But colonists also persecuted Native Americans,

taking land for themselves through decades of war and massacre. This is a tragic story in which mathematics plays little direct role. But this was the context in which Newtonianism travelled across the Pond.

The first establishments for higher education in America were based on those at Oxford and Cambridge. Harvard College was founded by the Puritan clergyman and Cambridge graduate John Harvard in 1636. Yale College then came along in 1701, created by a group of ten Christian ministers.

At the dawning of the eighteenth century, *Principia* had been published nearly fifteen years ago, but no copy had yet made it to America. Harvard tutor of mathematics and natural philosophy Thomas Robie had pieced together the ideas from journal papers published by the Royal Society which had built on Newton's ideas. A more direct route came via Robie's student, Isaac Greenwood, who travelled to London and learned mathematics from Newton's assistant. After returning to America, he began teaching Newtonian mechanics and became the first mathematics and philosophy professor at Harvard College.

By this point, Yale had become a distinct rival to Harvard and had managed to get hold of copies of both *Principia* (1713 edition) and *Opticks* (1704) from Newton via an intermediary who helped source scientific books from Europe. Yale's presidents, Thomas Clap and Ezra Stiles, quickly began teaching Newton's work at the university, even though they weren't mathematicians themselves.

Just as Newtonianism had spread in salons in Europe, a similar thing now happened in America. One particularly influential salon, founded in Philadelphia in 1727 by Benjamin Franklin, was called the Junto (from the Spanish word *junta*, meaning an assembly). Franklin's vision for it was that it was a club of 'mutual improvement', and it met on Friday evenings. It eventually evolved into the first learned society in America, the American Philosophical Society, in 1743. This society aimed to promote scholarly research and publications for both science and humanities and invited both domestic and international scholars to become members.

One of its members, David Rittenhouse, was an astronomer, a mathematician and one of the leading scientific voices in America. As a young boy, he had inherited a copy of *Principia* and became a lifelong advocate of Newtonianism. He believed that Newtonianism should be the foundation of higher mathematical learning in America and made this point in a speech to the American Philosophical Society in 1775. John Winthrop, a mathematician, astronomer, physicist and great-great-grandson of the founder of the Massachusetts Bay Colony, also saw the importance of Newton's work. He taught it at Harvard and established a laboratory there for testing the ideas.

Newtonianism and empiricism snowballed in America. After a slow start, more and more people began to know about Newton's work and applied it to other scientific disciplines, looking at how similar principles applied to electricity and magnetism. Benjamin Franklin, for example, famously conducted an experiment in which he created a battery from two charged bottles that was so powerful it could roast a turkey. This happened at the same time that Bassi was announcing the results of her own experiments with electricity.

The spread of Newtonianism to America was part of an eventual shift that would see the United States become a scientific powerhouse. The country would soon declare and fight for political independence from Britain but, scientifically and mathematically speaking, there would still be close ties. Harvard, Yale and other institutions were built in the European mould, and so when there were new mathematical developments, it was fairly straightforward for them to be adopted on both continents.

In China, things were different. Some Jesuit astronomers teaching at the Qing court translated Newton's work into Chinese in the eighteenth century. Europe was keen for its scientific ideas to journey eastwards as part of wider cultural diplomacy, ultimately aimed at spreading Christianity. However, China had its own mathematical traditions. Simply replacing Chinese mathematics with European mathematics was never going to happen. Instead, something far more interesting did.

10.   A Grand Synthesis

By the middle of the seventeenth century, the Great Qing dynasty was in full swing. The empire was one of the largest ever to exist, spreading from the Himalayas to Manchuria. It was multi-ethnic, with the Manchu ruling over a population that included Mongol people and Han people. People believed the emperor was the son of heaven put on Earth to rule the 'Middle Kingdom' – China literally means 'the centre of the world'. Those who ruled here ruled the entire universe.

China's intellectual influence was vast. The tributary states of East Asia, which include today's Korea, Japan, Thailand and Vietnam, were all heavily influenced by Chinese culture. Rulers would send diplomatic gifts and missions to China to show their respect to the emperor and in return received military protection and trading opportunities. The Confucian world order had been in place one way or another since the seventh century, but neo-Confucianism, a more secular and rationalist form of Confucianism, reached its height during the Qing dynasty. And this applied to mathematics too. Chinese mathematical traditions spread to the tributary states, erupting from the Forbidden City in Beijing, the mathematical and political centre of the Qing.

As the Renaissance in Europe kicked off a period of mathematical rediscovery, revision and revelation, mathematics in China also began to change. Its starting point was different, built upon the *I Ching* and *Nine Chapters*, but it too was heavily influenced by Arabic mathematics. Constant trade between people in China and the Middle East led to a regular exchange of ideas that helped refine the mathematical tools used in each location. The two mathematical traditions co-evolved.

China's relationship with Europe was a different animal. European Christian missionaries brought new (and old) ideas with them when they visited China in the sixteenth century and were, understandably, met with scepticism – so much so that one dispute resulted in a group of Jesuit mathematicians being sentenced to death.

However, the Qing dynasty would undergo a period of philosophical evolution and become more open to non-Confucian ideas. This led to a sort of 'grand synthesis', where Chinese mathematicians borrowed successful ideas and theories from Europe and the Arabian peninsula and merged them with their own. The result was a mathematics greater than the sum of its parts, but it was also a mathematics that was largely ignored by Europe. Most European intellectuals held prejudices that meant they incorrectly believed there was nothing to learn from Chinese mathematicians.

The grand synthesis was the biggest update in Chinese mathematics for a thousand years. It turned China into a highly technically proficient mathematical state. And it began with a seven-year-old boy dubbed the Kangxi emperor.

Eclipsed by an eclipse

By another count, when Xuanye was born in 1654 he would have been third in line to the throne. His father, the Shunzhi emperor, had two other, older male children, but their mothers were considered to be of a lower social standing. That meant that when the Shunzhi emperor died of smallpox in 1661, Xuanye, aged just seven, was put in charge and given the regal name of the Kangxi emperor.

At first, the Kangxi emperor, given his age, was a ruler in name only. His grandmother, the Empress Dowager Xiaozhuang, appointed four powerful men to run the empire as regents. After one of them died, the rest spent much of their time vying for political gain, with one having the other killed. It's remarkable, given this start, that the Kangxi emperor would go on to be the longest-reigning emperor in Chinese history, famed for ushering in a long period of stability and prosperity. In the mid-seventeenth century, he was still just a child, yet to seize full control of his empire. But it was against this backdrop that he would go on to develop a fascination with *xixue* (Western learning), particularly mathematics and astronomy.

Whenever a solar eclipse was due, the emperor and his regents tasked the astronomers at the Astronomical Bureau with calculating its exact timing and duration. The bureau was part of the Ministry of Rites, the agency in charge of setting the dates of important state rituals based on natural phenomena in the sky and on Earth. Neither solar nor lunar eclipses played much of a role in Chinese political astrology, but astronomers at the bureau realized they could serve a different purpose – eclipses were good indicators of the accuracy of a calendar.

Making accurate calendars was an important responsibility for emperors in China as part of their role as intermediaries between heaven and Earth. Calendrical accuracy boosted an emperor's credibility. It was generally accepted that good rulers knew the best times to perform rituals, so if something went awry it was a blot on their record. The bureau collected calendrical reference books written in Latin and Arabic which astronomers studied and translated into Manchu. However, the Chinese, Arab and European calendars all offered different approaches. A showdown was brewing.

At the time, the director of the Astronomical Bureau in Beijing was a Jesuit missionary named Johann Adam Schall von Bell. Many of the missionaries sent to China to spread Christianity also happened to have studied astronomy and mathematics. Schall had obtained his position after correctly predicting the time and duration of a solar eclipse under the Shunzhi emperor – an eclipse that local astronomers had got wrong using Arab and Chinese methods.

However, Chinese astronomers weren't exactly happy about a Jesuit being in charge. Confucian scholars were particularly concerned, worried that having a foreigner guide calendrical calculations could lead to the abandonment of Confucian principles and moral collapse. Schall's work began to come under scrutiny, especially from a most unusual character named Yang Guangxian.

Yang wasn't your typical astronomer. He had little in the way of a mathematics background but had instead learned astrology and fortune-telling while exiled from Beijing for blackmailing Ming officials. When he returned to Beijing after the fall of the Ming dynasty, he started calling himself an astronomer but found that all the highest positions were occupied by Jesuits. In an attempt to bring down

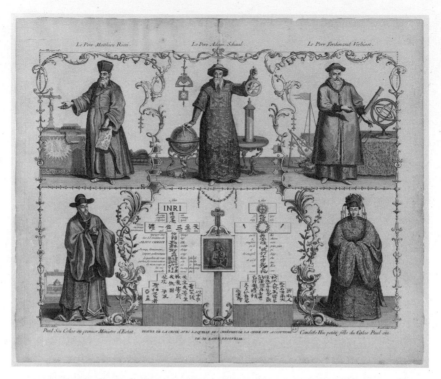

Jesuits in China. From Jean-Baptiste Du Halde, *A Geographical, Historical, Chronological, Political and Physical Description of the Empire of China and Chinese Tartary*, Paris, 1735.

Schall, he publicly accused him of choosing an inauspicious day for the funeral of Shunzhi's fourth son, who had died at the age of three months. He followed this up with a pamphlet attacking the Western calendar and Christianity. The Jesuits, Yang argued, had plotted rebellion, propagated a heretical religion and disseminated erroneous knowledge of astronomy.

Qing government officials took the claims seriously. Another astronomer, Wu Mingxuan, publicly backed Yang, saying that there were defects in Schall's method. Wu's family had Islamic ancestry and served successive dynasties as astronomical experts for over a millennium. However, he had now fallen out of favour, after a method he was using was deemed inferior to one known by the Jesuit missionaries, leaving him without a job.

The tide quickly turned on Schall. Government officials were already sceptical of the Jesuits and so put him on trial. He suffered a stroke during the proceedings and so his friend the Flemish Jesuit priest and proficient astronomer Ferdinand Verbiest helped to defend his methods.

In April 1665, government officials found Schall guilty of creating an erroneous calendar and choosing the wrong date for an important state ritual. They imprisoned him along with Verbiest and other colleagues at the Astronomical Bureau – three Jesuits and five Han Christian converts in total. Schall was sentenced to death by being cut up into pieces. His colleagues were given a hundred lashes and expelled from their posts at the bureau. Yang, on the other hand, was made an imperial astronomer by the Kangxi emperor and his regents, and Wu got back his position at the bureau.

But by sheer luck, Schall was spared. A massive earthquake destroyed the execution site and a rare meteor was observed in the sky. The superstitious regents of the Kangxi emperor viewed this as an omen that the execution should not take place. The Jesuit missionaries were released and exiled to Canton, in southern China. The Han astronomers had no such luck, however, and were beheaded.

Yang performed poorly while in charge of the bureau. He claimed to have invented a new method for calendrical calculations, which he called the Calendar of Timely Modelling, but in reality he had just rebranded the ancient Great Concordance system. He ordered the officials at the Astronomical Bureau to stick to these methods. In 1666 and 1667 this led to mistakes, for example the prediction that there would be two spring and two autumn equinoxes each year. By 1668, suspicion of Yang's abilities was rising and Chinese court officials were worried about the accuracy of the calendar his team had produced for the following year.

Schall had since died, but Verbiest and the other exiled Jesuits were invited back to Beijing to help the Board of the Rites translate some letters from the Dutch governor of Batavia (today's Jakarta). While there, government officials decided to run the calculations by him, and he spotted an error. Yang and his team had only used the hours of sunrise and sunset for the capital of Batavia, but Verbiest used the hours of sunrise and sunset for the capital of each province, and by doing so he was able to see why it had all gone wrong.

On hearing of the potential problem, the Kangxi emperor agreed with Verbiest and Yang that they would each demonstrate their methods by performing three tests: to predict the length of shadow produced by a sundial; to predict the position of the sun and some planets on a given date; and to predict the time and duration of an upcoming lunar eclipse. 'He who errs the most shall be judged to have the most mistaken mathematics,' said Verbiest.[1]

The then fourteen-year-old emperor summoned the pair to his palace observatory, where the tests would take place over three days in front of him and a noisy crowd. On each occasion, Verbiest was triumphant and was met with applause. The demonstration convinced those present of the utility of European mathematics. The emperor recalled later that 'There was no one who knew the method' of the actual calculations. He was inspired by this; as he later said to his sons, 'I realized that if I didn't know it myself, I could not judge true from false. So I was eagerly determined to study mathematics.'[2] Verbiest was made director of the Imperial Observatory and the Board of Mathematics and became Kangxi's mathematics tutor.

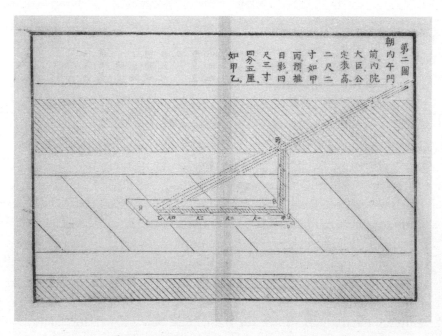

A gnomon that was set up within the court on 28 December 1668.

Tapestry of a Chinese emperor with Jesuit priests.
Made under the direction of Philippe Béhagle between 1690 and 1710.

West meets East

The Kangxi emperor took maths lessons daily. Starting from Euclidean geometry, Verbiest and other Jesuit missionaries presented arithmetic, algebra, trigonometric tables, logic and the European methods for predicting the paths of the moon and the stars. The emperor incorporated this new knowledge into the Chinese techniques he already knew, often using his abacus to make calculations faster than the missionaries.

The first six books of the *Elements* had already been translated into Chinese by Italian Jesuit Matteo Ricci and his Chinese scholar friend Paul Xu Guangqi, between 1607 and 1610. The publication led to the creation of a new term, *jihe*, to mean 'geometry', and Chinese mathematical books began to adopt it. But many had been sceptical of the work and suspected that Ricci was hiding something by not translating the remaining seven books. The Kangxi emperor's enthusiasm for European mathematics changed these views, kicking off a long and careful review of the two mathematical traditions.

One of the people tasked with this was Mei Wending. Mei was from a talented mathematical family and had encountered European astronomy as a result of Schall's ordeal in prison. The news had spread quickly throughout the country and when Mei heard of it he wanted to understand the mathematics at the heart of it for himself.

He actively investigated mathematical texts, both Chinese and European, in the hope of determining the advantages and disadvantages of each. For example, when it came to simultaneous equations, he strongly believed that European mathematics lagged behind traditional Chinese techniques, writing to a friend, 'I am disgusted by those Western missionaries who exclude traditional Chinese mathematics and therefore I wrote this book about which even Matteo Ricci could not possibly say a bad word.'[3] This book was *On Simultaneous Linear Equations*, published in 1672, and it clearly outlined how much more developed the Chinese study of simultaneous equations was compared to that in Europe.

Chinese mathematics already had an extremely advanced form of algebra by the fourteenth century, and much of it focused on practical commercial applications. Imagine a situation where you had 125 copper coins with which to purchase three bundles of silk and two bunches of cotton. You know that the silk always costs thirty more copper coins than cotton, so you need to work out the highest price you can afford to pay for each. In our notation, we write:

$$3x + 2y = 125$$
$$x - y = 30$$

where x is the cost of the silk and y is the cost of the cotton. We write this with $+$, $-$ and $=$ signs, but at the time neither Chinese nor Jesuit mathematicians would have written the problem as an equation. Both could have described it using words, but Chinese mathematicians also had another method, involving counting rods.

As we've mentioned before, to solve simultaneous equations like these, Chinese mathematicians used counting rods coloured black for negative and red for positive. They then had various methods for moving the physical counting rods around to efficiently solve the equations.*

In the European style that the Jesuit mathematicians in China were promoting, solving the problem was often more arduous. They would manipulate the terms to put the problem into standard forms to which a particular rule applied, much as al-Khwārizmī had first described. For equations involving higher powers such as x^2 or x^3, European methods became even more cumbersome. They relied on using bespoke words for the various powers of the unknowns, such as *radix*, *zens* and *cubus*, but this made these polynomial equations difficult to express. Chinese mathematicians had elegant ways to use counting rods to deal with equations of this form.

Take $x^4 + 2x^3 - 13x^2 + x - 265 = 0$, for example. This could be represented using rod numerals by placing the coefficients vertically, from the highest powers to the lowest.

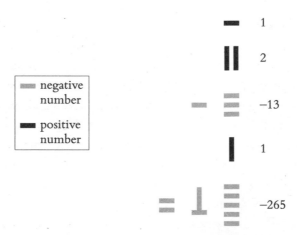

Chinese mathematicians then utilized a technique called the method of the celestial unknown to solve the equation. There was at this time no comparable method of solving polynomial equations such as this in

---

* $x = 37$, $y = 7$, for the pair above, in case you were wondering.

Europe. One did eventually appear in 1819 and is often given the name 'Horner's method', after William George Horner. In China, thirteenth-century mathematician Qin Jiushao was one of the first to hit upon the approach, in 1247, although mathematicians in the Middle East came up with a similar idea, both before and after Qin, including Nasir al-Din al-Tusi of House of Wisdom and Tusi-couple fame, who wrote about a comparable method in 1265.

No matter where these methods first originated, Mei was right to believe that Chinese mathematics was not inferior to European and would spend the next twenty years working out how best to mix the two. In one of his astronomical works, *The Doubt concerning the Study of Astronomy*, he wrote a dialogue between two scholars, one who doesn't know anything about astronomy and the other a specialist. Perhaps a little dangerously, the ignorant scholar was modelled after seventeenth-century Confucian scholar Li Guangdi and the specialist resembled Mei himself.

In the first chapter, the two discuss what they termed the Chinese, Western and Muslim astronomical systems. In China, the Triple Concordance System was still prevalent, but by the end of the fifteenth century many knew it was inaccurate, and an updated version called the Season Granting Calendar was gaining ground. However, even this updated version did not accurately predict lunar eclipses. The two characters concluded that Western astronomy was more accurate but that Muslim and Western astronomy had the same origin. The main difference, they said, was that Western astronomy had been improved using observational data taken with telescopes, yet it was no more advanced. Mei concluded, 'Number and principle are united', and 'China and the West do not differ.'⁴ Mathematical knowledge, as he saw it, was universal.

As he worked his way through, Mei added to and updated this knowledge. The Gougu Theorem, for example, had been untouched since the third century. Liu Hui, as we saw, covered it in his textbook, but no proof had been given in any other mathematical books in China. Mei presented two new proofs.

One of the reasons Mei was able to do so much to update and challenge old ideas was that he was part of a new movement that believed in *gewu qiungli* ('the investigation of things'). Unlike many of the Confucian scholars before him, who simply relied upon the word of Confucius and other sages, Mei believed that by studying the world around him he could uncover general principles about how it worked. The rise of evidential research was good news for mathematics. Mei started to assemble nine treatises under the title the *Integration of Chinese and Western Mathematics*. He learned from Jesuit mathematicians but always treated what he learned with a critical eye, aware that throughout Western mathematics some ideas were more advanced, some were inferior and some were just different. It eventually became his goal to unite the two mathematical systems.

In 1703, when he was seventy years old, Mei's work was noticed by the Kangxi emperor and he was summoned to the Forbidden City. He presented some of his ideas, among them his proofs of Gougu and a range of methods to survey land. After the audience with the emperor, Mei composed a poem, ending one line with 'Ancient and modern, Chinese and Western are all consistent.' This audience led to the appointment of his grandson Mei Juechen, as a court mathematician in 1712 and to the establishment of the Academy of Mathematics in 1713, which was modelled after the French Academy of Sciences.

Best of three worlds

Mei Juecheng and other mathematicians trained by Mei Wending continued to work on fusing Chinese, Arab and European mathematics. They reviewed Mei Wending's work on the motion of the moon and found that it could provide more accurate predictions of lunar eclipses than any previously known methods. The key was to use the best of all available astronomical data, no matter where it came from.

The Board of Mathematics took a similar approach. Under the direction of the Kangxi emperor's son Prince Yunzhi they published a nearly 5,000-page book that combined the various different mathematical traditions. It was called the *Essence of Numbers and Their Principles Imperially Composed*. This ambitious reference work epitomized the grand synthesis of mathematics, starting from ancient

Chinese astronomy and *Nine Chapters*, taking a tour to the *Elements* and the mathematics brought by the Jesuits, before reaching Mei Wending's examinations and additions.

Wang Zhenyi then took up the baton. She was born in 1768 into a scholarly family in Suzhou prefecture and during her childhood spent many hours reading in her late grandfather's library. She also practised martial arts while galloping on horseback and learned from the wife of a Mongolian general how to hit targets with a bow and arrow while riding. Inspired by her grandmother's love of writing poems, she wrote *In Praise of Manly Women*, in which she declared, 'I am for extensive travelling and reading, even greater than man's is my ambition.'

She directed her ambition towards mathematics and astronomy. Self-taught, she was particularly influenced by one of Mei's books, *Principles of Calculation*. She wrote her own version in simpler language that non-specialists could understand. It was quite the task: 'There were times that I had to put down my pen and sighed. But I love the subject, I do not give up.'[5] She published her five-volume book, *The Simple Principles of Calculation*, at the age of twenty-four.

Wang's writings on astronomy were particularly provocative. Although many astronomers in China disliked the Western calendar, she urged them to adopt it. 'What counts is the usefulness, no matter whether it is Chinese or Western,' she wrote.[6] In her essays she explained how to determine equinoxes and demonstrated how to calculate them. She traced the orbits of celestial bodies as well as sketching how they rotated on their axes. And she described how both lunar and solar eclipses happen. Her books had more detailed observations and explanations than any others in China at that time.

Like Laura Bassi, Wang set up experiments outside. A crystal lamp on a cord dangling from a garden pavilion represented the sun. A round table outside represented Earth. On one side of the table, Wang put a round mirror to represent the moon, then moved all the objects to see the relationships between them. She wrote an explanation of the solar eclipse based on her experiments.

Wang was a prolific writer and disseminator of mathematical knowledge. She was also a strong advocate for equal rights for men and women. Among Wang's manuscripts were criticisms of women's place in Confucian society. She lamented that, in discussions about learning and the sciences, women weren't even part of the conversation; instead, they were expected to 'only do cooking and sewing'. She summed this up in a short poem that was found in her unpublished notes after she died.

> It's made to believe
> Women are same as Men;
> Are you not convinced
> Daughters can also be heroic?[7]

Wang was one of few women in the world of mathematics at the time. The approach she encapsulated of analysing and then combining the best ideas around the world to create a rich body of scientific knowledge feels palpably modern and led to Chinese mathematics becoming arguably the most advanced of the time. Very little similar synthesis took place in Europe. The Catholic Church was so opposed to Confucian traditions that they rejected Chinese knowledge in its entirety, whatever the subject matter. Nor did learned European academies and societies make much attempt to understand Chinese mathematics. Mathematics was intertwined with religion, ideology and identity. It wasn't just about mathematical tools but was a way of life.

One notable exception was one of the 'King's mathematicians', Joachim Bouvet. He was a Jesuit missionary sent to Beijing in 1685 as part of King Louis XIV of France's delegation. He was invited to lecture at the court there and during his stay made astronomical observations which both the Chinese emperor and the French Academy of Sciences made use of. Bouvet learned the Chinese alphabet and believed that the characters held important symbolic meanings. Some missionaries, including Bouvet, believed that many of the Chinese classical texts foreshadowed important Christian events, including the birth of Jesus, so its proponents took a very different view of Chinese knowledge than did much of the Catholic Church. Bouvet believed that the *I Ching* was the oldest book in existence and could reveal mysteries about Christianity. It was in this book that he found the hexagrams he would show to Leibniz. Ultimately, however, his focus on Chinese mathematics would not shift the dial in Europe.

It's fascinating to wonder how the history of mathematics would be different today had European mathematics engaged in a similar synthesis of ideas to that which took place in China. As we have seen again and again, there is no one true mathematics. Instead, it is an ever-evolving body of knowledge affected by culture, location and time. And times were certainly beginning to change.

11. The Mathematical Mermaid

Paris, 1888. The entries were in for the illustrious Prix Bordin, a mathematical prize set by the French Academy of Sciences. In order to prevent reputations from influencing the judges, each entry was marked only by a phrase. As they worked their way through the entries, the judges saw that one stood head and shoulders above the rest. It described a solution to a mathematical problem that had remained unsolved for more than a hundred years and had left Euler and Lagrange, two prolific mathematicians, scratching their heads.

The author of this entry was identified only by the phrase: 'Say what you know, do what you must, and whatever will be, will be.' It was the mathematical nom de plume of someone who had spent their life experiencing discrimination, setbacks and personal tragedy but had now finally made it. Perhaps it was this mentality that had got them to this point – along with unwavering, dogged determination.

'Gentlemen,' said the astronomer and president of the Academy, Jules Janssen, beginning the prize announcement, 'among the crowns that we are about to bestow, there is one of the most beautiful and difficult to obtain which will be placed on a feminine brow.'[1] The winner of the Prix Bordin that year was Sophie Kowalevski, the world's first female mathematics professor.

This moment had been coming. Over the previous decades and centuries women had been making their mark on mathematics, challenging views and norms. Ever since the seventeenth century, mathematicians in Europe had slowly started to shift from being amateurs to professionals, and professorships were a large part of that. But professorships were also political tools, given to those in the club and not to those outside the establishment. Somehow, Kowalevski had managed to break into this world – but her entry would come at a cost.

A little more liberal

The best place to begin this story is in the region that would eventually become Italy. The oldest record we have of a female Ph.D. is from 1678. Of Albanian heritage, Elena Lucrezia Cornaro Piscopia was a multilingual philosopher who lived in Venice. She was born into a noble family and as a child became fluent in Latin, Greek, Hebrew, Spanish, French and Arabic. Somehow, she also found time to master the harpsichord, clavichord, harp and violin. Her tutor Carlo Rinaldini, a professor of philosophy at the University of Padua, also taught her mathematics. She took to it so quickly that Rinaldini compiled a customized geometry textbook for her, published in 1668. By her teens Cornaro had already surpassed the knowledge of an average undergraduate degree. Her father – an influential cardinal – championed his daughter's talents and suggested she apply for a doctorate.

Officials in the Roman Catholic Church initially denied the request because she was a woman, but the bishop of Padua eventually sided with her father. Cornaro obtained her doctorate degree in 1678 at the age of thirty-two, after passing the oral examination at the University of Padua in 1678. Rather than taking up a teaching position at the university, Cornaro decided to prioritize a life of charity over one of mathematics. She spent her time helping the poor. However, she was plagued with ill health and died at just thirty-eight, probably of tuberculosis.

The title of first female mathematics professor nearly went to another woman living in Italy – Maria Gaetana Agnesi. Born in Milan in 1718, Agnesi, like Cornaro, had a talent for languages. She spoke Italian and French by the age of five and had learned Greek, Hebrew, Spanish, German and Latin by her eleventh birthday. At twelve, she began to suffer from a mysterious and persistent condition. At the time the suspected cause was intense study, but it was probably some form of mental illness. As a result, on the advice of her physician, Agnesi spent time at a country villa in Masciago, twenty-five kilometres north of Milan. The more relaxed rural life enabled her to enjoy horse riding and dancing, yet she was still haunted by violent nervous attacks and she attempted suicide more than once.

By 1733, she had recovered and was back home in Milan, resuming her various areas of study. She became proficient in metaphysics, moral philosophy and mathematics, mastering calculus. She collected a vast library of around four hundred books and wrote her own, *Analytical Institutions*, an introduction to mathematics. She was meticulous in how the book should be printed. The publisher, Richini, installed a printing press on the ground floor of Agnesi's house so that she could supervise the details. She was especially careful with the symbols of differential and integral calculus, worried that the non-mathematician typographers might make mistakes.

Title page of Agnesi's *Analytical Institutions*, which was printed at her home.

Two years after the book's publication, Agnesi's reputation had grown and her name became known to Pope Benedict XIV. The pope believed that women's education was at the core of enlightened Catholic culture, writing, 'From ancient times, Bologna has extended public positions to persons of your [Agnesi's] sex. It would seem appropriate to continue this honourable tradition.'[2]

In 1750, Agnesi was offered the honorary chair of mathematics and natural philosophy at the University of Bologna, which was known as the new 'paradise of women'. It shared this liberal and inclusive atmosphere with her hometown, Milan. But her illness came back again, and her doctor advised against accepting the position. She would instead go on to study theology and devote her life to charity, much as Cornaro had done.

Equations on the wall

Sophie Korvin-Krukovskaya was a child prodigy. Born in Tsarist Moscow in 1850, she became determined to break the barriers stopping women from pursuing a career in mathematics. Her father, Vasily, was impossibly strict and a firm believer in the patriarchal norms present in Russia at that time. He thought that women needed only enough schooling to participate in fine society, and no more. Having a learned woman in the family, Vasily believed, would bring shame on all other members. So Sophie's first contact with mathematics came not from her father but from her uncle Pyotr, who often visited their house, and more frequently still after his wife was murdered by their servants.

Their bond grew strong during these visits. Pyotr would often talk with her and, as an educated man who saw a spark in his young niece, he would discuss mathematical concepts, for example telling her about asymptotes, a line that gets infinitely close to a curve. As Sophie would later recall in her memoir, she didn't at first understand these ideas, but they fascinated her. 'The meaning of these concepts I naturally could not yet grasp, but they acted on my imagination, instilling in me a reverence for mathematics as an exalted and mysterious science, which opens up to its initiates a new world of wonders,' she wrote.[3] Her uncle had piqued her interest, and she would soon come face to face with more mathematics.

When she was eight, her father retired from his military position as head of the Moscow Artillery and the family moved to a new house in the country. The house in Palibino had just been renovated and was still in need of wallpaper. After decorating several of the rooms, the family ordered more wallpaper from St Petersburg, but it never arrived, due to 'rustic laxity and characteristic Russian inertia', as Sophie would later put it. After careful reconsideration, the family decided that the nursery would not be as well decorated as the rest of the house. Embracing an upcycling spirit, they resolved the situation using Vasily's old mathematical notes. To prepare for his job as an army officer, he had studied differential and integral calculus. Aged eleven, Sophie felt a strange attraction to the mysterious equations on the wall. She had no idea what they meant, but she had a strong feeling that it must signify something 'very wise and interesting'.[4] It seemed as if the symbols were popping out of the wall to speak to her.

Although she didn't initially have the chance to take classes in mathematics, one tutor, Yosif Malevich, did teach her some basic arithmetic, geometry and algebra. It was clear, even then, that she had a gift for the subject. Malevich gave her a copy of an algebra text-book, which she read avidly from cover to cover. However, her father still opposed the education of women and so stopped the tuition. His daughter would not be deterred, studying her algebra textbook under the cover of darkness, before a stroke of good fortune brought mathematics into her life once more.

An academic named Nikolai Tyrtov was a neighbour and family friend, and when he published a new textbook on physics he brought a copy to the house. She was mesmerized by it. The optics section, which analysed the physics of light, particularly caught her attention. Sines, cosines and tangents – the functions of trigonometry – punctuated the pages. The formulas did not look like anything that she had ever come across before. Slowly but surely, she made her way through the book. When she related her progress to Tyrtov, he was impressed and believed her to be a prodigy. He spoke with her father, persuading him to give his daughter a proper mathematical education. In Tyrtov's estimation, she was the 'new Pascal'.

Vasily finally relented and hired a new teacher in 1867. Sophie was given a broad education in mathematics by Alexander Strannolyubsky, the author of Russia's first manual for teaching and learning algebra. As her mathematical ability matured, she became increasingly obsessed with the subject. She became desperate to free herself from her father, who continued to have strong prejudices against learned women and would not agree to her pursuing a mathematical career.

Sophie was eighteen when the opportunity to escape her father finally arose.

In the late 1860s, nihilism, a new philosophy of radical socialism, was gaining prominence. The movement advocated the total rejection of all existing authority and instead encouraged the study of, and belief in, science. The younger generation came under the spell of nihilism, seeing it as a means to challenge orthodoxy. Sophie's older sister, Anna, was seduced by this new philosophy and decided to leave Russia. She was looking for a chance to move but, as a single woman, she needed her father's signature in her passport to travel abroad, and he refused to give it. So she sought another option. The quickest way out was to enter into a 'fictitious' marriage, known among the nihilists as a 'white marriage'. By marrying someone who was sympathetic to her cause she would be able to move abroad and attend a college that accepted women.

She set about gathering suitors, one of whom seemed particularly suitable – a young, radical and well-travelled twenty-six-year-old named Vladimir Kovalevsky. He had been to London and on his return busied himself with translating Charles Darwin's most recent book, *The Variation of Animals and Plants under Domestication*, into Russian. He was so impressed with Darwin's ideas and worked so rapidly that his translation came out before the English publication. Although Vladimir believed that both sisters should be freed from their oppressive family, he became infatuated with Sophie, and not Anna, and so proposed a white marriage to her.

But her father would not agree. Sophie was only eighteen, and Vladimir was eight years her senior. Inspired by Dostoevsky's novels, she made a scene, locking herself in Vladimir's apartment and declaring to her parents that she would not come out until her father agreed to her marriage. Finally, he relented. Sophie Korvin-Krukovskaya became Sophie Kowalevski and, after her marriage, the world unfurled before her. She studied in St Petersburg and Heidelberg before eventually moving to Berlin. It was here that she met Karl Weierstrass, the German mathematician who would be instrumental in her rise to fame.

Sophie Kowalevski.

Productive as a pair

When Kowalevski and Weierstrass first met, they already knew one another by reputation. Weierstrass was a world-famous mathematician, and praise for Kowalevski's talents had reached him from her professors in Heidelberg. To test if the rumours were true, Weierstrass sent Kowalevski a series of mathematical problems normally reserved for the most experienced (male) students. She solved them with aplomb. Weierstrass was impressed. Kowalevski really had had only minimal mathematical training compared to the rest of his students, but her ability was so evident he took her under his wing. In Kowalevski's words, this decision had 'the deepest possible influence on my entire career in mathematics'.[5]

Under Weierstrass's tutelage, Kowalevski's first major mathematical discovery had to do with the shape of Saturn's rings. Almost a century before, French polymath Pierre de Laplace had suggested that Saturn had a large number of solid rings, but no one could describe an exact shape for them that would fit astronomical observations. Kowalevski suggested a new approach: the rings could be made out of fluid rather than solid material. Using infinite series – similar to those studied separately by Mādhava, Leibniz and Newton – she showed that if the rings were made of fluid they would be egg-shaped rather than perfectly symmetric ovals, as had previously been thought. Although astronomers later found that Saturn's rings were in fact made almost entirely of small pieces of ice, the mathematical methods she developed would find many other applications in the field of geometry.

Kowalevski quickly amassed a body of original research. If she had been a man, it would have been more than enough to earn a doctorate, but at the University of Berlin women were effectively barred from gaining one. One requirement for obtaining a Ph.D. there was that doctoral candidates had to undergo a *viva voce*, an oral defence of their work in front of a group of experts. Women weren't allowed to participate. Eventually, Weierstrass decided to package up three of Kowalevski's mathematical papers, including the one on Saturn's rings, and send them to the University of Göttingen. In Göttingen, doctoral candidates could graduate *in absentia*, that is, without turning up to defend their theses – providing it was considered good enough. Göttingen thought Kowalevski's work certainly was, and so in the summer of 1874 she finally obtained her doctorate.

Two decades later, a survey of a hundred eminent professors in Germany outlined their views on women in academia. Just under half were positive. For the time, this was progress. But a third of respondents still thought women should not study at all. Women often had to come up with inventive ways in which to sidestep prejudice.

Take the case of Monsieur LeBlanc, a name written at the bottom of a letter received by Sardinian mathematician Joseph-Louis Lagrange. Lagrange was a Newtonian and a founding professor of mathematical analysis at the École Polytechnique, an engineering school on the outskirts of Paris founded in 1794. Monsieur LeBlanc seemed to be a student taking Lagrange's class and was writing to ask some questions. Even though Lagrange didn't recognize the name, he saw enough talent demonstrated in the letter to take it seriously. He wrote back and subsequently the pair became pen pals.

Monsieur LeBlanc also wrote to Adrien-Marie Legendre, another member of the mathematics faculty at the École Polytechnique, with an attempt at proving Fermat's last theorem. Fermat's last theorem had long held mythical status among mathematicians because of its deceptive simplicity. There are whole numbers that neatly slot into the equation $x^2 + y^2 = z^2$, such as 3-4-5 and 5-12-13, and these are known as Pythagorean triples (though perhaps more accurately called Gougu triples). You might think that similar numbers would exist for $x^3 + y^3 = z^3$ or $x^{14} + y^{14} = z^{14}$, but in fact, as Fermat conjectured, there are no whole numbers that satisfy the equation $x^n + y^n = z^n$, where n is greater than 2. He famously scribbled in one of his notebooks, 'I have a truly marvellous demonstration of this proposition which this margin is too narrow to contain.'[6] Unfortunately, nobody ever found that proof – if it ever even existed.

Monsieur LeBlanc worked on Fermat's last theorem, especially the case in which $n = p$, where p is a number that does not divide xyz without leaving a remainder, and proved it for every prime p less than 100. He clearly impressed Lagrange and Legendre, so they wanted to meet LeBlanc. They were shocked, however, to discover that Monsieur LeBlanc was in fact a woman, Sophie Germain.

Germain was the daughter of a wealthy merchant and had grown up reading books in the family library. Her parents had strongly objected to her studying mathematics and used to confiscate warm clothing and lights from her room to prevent her studying it at night.

Defiant, she would wrap herself in a sheet and study by candlelight. Eventually, her parents realized that they had lost the battle and allowed her to study further.

Specialized schools such as the École Polytechnique in Paris, where she lived, did not accept female students at the time. Yet she managed to obtain the notes of Lagrange's lecture notes by asking him personally, and studied mathematics on her own. When Lagrange found out who she really was, he was so impressed by her knowledge that they continued to discuss maths and he became her tutor. She went through a similar series of events with German mathematician Carl Friedrich Gauss, who was also interested in Fermat's last theorem. 'Unfortunately, the depth of my intellect does not equal the voracity of my appetite,' wrote M. LeBlanc, 'and I feel a kind of temerity in troubling a man of genius when I have no other claim to his attention than an admiration necessarily shared by all his readers.'[7] In contrast to the humbleness of her letter, Germain's claim to Gauss was remarkable: she said she had proved Fermat's last theorem in its entirety.

Would Germain be the one to crack the problem? Gauss wrote back, 'I am delighted that arithmetic has found in you so able a friend.' However, he went on to say that though her proof was good, it only worked for a particular case and wouldn't work for other numbers.[8] She had, though, made some progress. Her proof worked only in the special case where $n = p - 1$, where p is a prime number of the form $p = 8k + 7$, for some whole number k. Nonetheless, the methods that she discovered would form the basis of attempts to crack Fermat's last theorem for hundreds of years to come. (It would be solved in 1995.)

A particular type of prime number turned up in her explorations of Fermat's last theorem that now bears her name. A prime number p is a Germain prime if $2p + 1$ is also prime. For example, 11 is a Germain prime as $2 \times 11 + 1 = 23$ is also a prime number. Germain primes turn out to be important in cryptography; the resistance to attack of some encryption systems hinges on choosing a good prime number, and Germain primes are a particularly good choice.

Germain kept working on mathematics. When she learned that Napoleon was about to invade Prussia in 1870, she feared that Gauss, who was in Berlin, might be killed during the invasion. Germain asked her friend General Joseph-Marie Pernety to go and protect Gauss. Pernety told Gauss that he owed his life to Mademoiselle Germain, and this resulted in Germain revealing her true name in her next letter to Gauss. On 20 February 1807, she wrote, 'fearing the ridicule attached to a female scholar, I have previously taken the name of LeBlanc in communicating to you those notes'. Gauss replied on 30 April 1807, 'how can I describe my astonishment and admiration on seeing my esteemed correspondent M. LeBlanc metamorphosed into this celebrated person'. Gauss was empathetic to her situation and went on to write, 'when a woman, because of her sex, our customs and prejudices, encounters infinitely more obstacles than men in familiarising herself with [number theory's] knotty problems, yet overcomes these fetters and penetrates that which is most hidden, she doubtless has the most noble courage, extraordinary talent and superior genius'.

Meeting the mermaid

Sophie Kowalevski's husband, Vladimir, completed his doctoral research at the University of Jena while she was in Berlin and became a hippopotamus specialist. But work for such esoteric experts on sub-Saharan megafauna was hard to come by in central Europe, and the pair spent much of the time completely broke. She applied for jobs in Germany, but no university would even consider a female candidate. Disappointed, she returned to Russia and decided to refocus her energy on her marriage.

Up to this point, the pair had mostly lived apart, but as they became geographically closer, they became romantically closer too. They started a real-estate business, hoping to make enough money to become financially independent and pursue their scholarly interests. After a while, this fell by the wayside and the pair began to turn to other work. Sophie Kowalevski transformed herself into a writer of fiction, theatre reviews and popular-science reports for newspapers. However, without a job in mathematics, she had little choice but to halt her scientific work. The couple had a baby, known by her nickname, Foufie, but her husband continued to struggle to find work and they were forced to declare bankruptcy.

In January 1880, Pafnuty Chebyshev, an eminent Russian mathematician, asked Sophie Kowalevski to give a talk in St Petersburg. She hadn't given up on her dream of becoming a professional mathematician, so she jumped at the opportunity. She translated an unpublished part of her thesis from German into Russian overnight and delivered the lecture the next day. This proved to be a turning point: the moment her dreams came within touching distance.

Listening to her presentation, Gösta Mittag-Leffler, a former student of Weierstrass and now a professor of mathematics at the University of Helsinki, was enormously impressed, and determined to find her a job in his native Sweden. He was due to take up a position as a professor of mathematics at the University College of Stockholm the following year and hoped that he might be able to use his influence to her benefit. Kowalevski and Gösta began to correspond and in 1881 she wrote that she had been obsessed with a problem that was beautiful and elusive, one she called the 'mathematical mermaid'.

The problem is one that ballet dancers solve intuitively all the time. When they launch into a pirouette, spinning on one foot, they can alter the way they move by adjusting the position of their arms or their other leg. As they whirl around, they can speed up or slow down with the slightest of bodily tweaks. The variables – shape, acceleration and speed – are easy for them to comprehend. A tweak in one changes the other. By mastering the relationships between the variables, ballet dancers can time their rotations to perfection. Mathematicians, however, had no such luck. Even a spinning top that wasn't completely round couldn't be described mathematically. It seemed too random, and too difficult to express in an equation.

Kowalevski set her sights on determining the mathematics of a spinning top. She wrote, 'This research struck me as so interesting and engaging that for the time being I have forgotten everything else and given myself over to this work with all the fervour of which I am capable.'[9] Desperate to return to research in maths, she moved back to Berlin, now with her daughter, who was just a few years old. There had been difficulties in her marriage for some time and she did not tell her husband about the move.

She rekindled her work with Weierstrass and spent two further years in Berlin. Meanwhile, Gösta married a woman from a very wealthy family whose dowry allowed him to launch a mathematical journal, *Acta Mathematica*, in 1882 and build an elegant villa in the north of Stockholm. He founded a private library of mathematics and his influence in the mathematical community spread.

In the spring of 1883, Sophie Kowalevski's life was struck by tragedy. In her absence, Vladimir had tipped into depression and he took his own life. Kowalevski was devastated. The man who had helped her most was dead and her feelings of guilt led to attempts to starve herself to death.

Gösta got in touch at just the right time. He now had enough influence to secure Kowalevski a teaching position at the University of Stockholm, and this gave her a reason to live. She picked herself up, accepted the position and relaunched herself into her lifelong dream of becoming a professional mathematician. It was the beginning of a new chapter in her life, but it was far from easy. As she was a woman, she received no salary. Instead, she had to personally collect tuition fees from her students. She elicited mixed responses from colleagues and observers. Playwright August Strindberg claimed that 'a female professor is a pernicious and unpleasant phenomenon – even, one might say, a monstrosity'.[10] Reporting her arrival, one Stockholm newspaper described her as 'a princess of science'. Her response: 'Look at that! I've been made into a princess! It would have been better if they had given me a salary.'[11]

She continued to teach in Stockholm, leaving her daughter in Russia with her brother-in-law. In May 1884, she wrote: 'I think it would be an unforgivable weakness on my part if I allowed myself to be influenced even slightly by the wish to *appear* a good mother in the eyes of the old biddies of Stockholm.'[12] Her mathematical peers were soon won over by her abilities. In 1883, she published mathematical papers on the properties of light that attracted much admiration. Students at the university also thought she was an excellent teacher. So, in July 1884, she was appointed as a professor of mathematics at the University College of Stockholm. In doing so, Kowalevski made history.

Turning pro

As a professor, Kowalevski had the freedom to devote her time to the spinning top, her mathematical mermaid. Her work built on that of two mathematicians who had attempted to crack the problem a century earlier – and had made some progress. Back in 1758, the Swiss mathematician Leonhard Euler came up with the equations that describe the situation where the point at which the top spins is the same as its centre of gravity (left in the diagram below). With these equations, given the starting speed and position, it would be possible to work out where the top would have rotated at a later point in time. Joseph-Louis Lagrange then extended the equations to encompass any symmetrical spinning top. The remaining task was to extend the equations further to the case where the spinning top isn't symmetrical. But that proved too difficult for either of them.

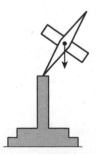

Three examples of spinning tops. The movement of a rotating object depends on its shape. The object on the left has a symmetrical ornament on top; the centre of gravity is on the axis of movement. The object in the middle also has a symmetrical ornament, but here the centre of gravity is not on the axis but at the centre of the ornament. The object on the right has an asymmetric ornament due to the small weight placed on one side. The centre of gravity does not align with the axis, nor is it contained within the object itself.

Over the course of a year, Kowalevski devoted all her research time towards solving the problem of the mathematical mermaid. Her breakthrough came when she tried to describe the path that a point on a spinning top takes and stumbled upon a tool called 'theta functions'. Although theta functions were not new to mathematics, she realized she could use them to simplify the problem. Euler and Lagrange's work could accommodate only one changing part of the equation or variable, representing the way a symmetrical object spins. But when you move to the non-symmetrical case, you need two

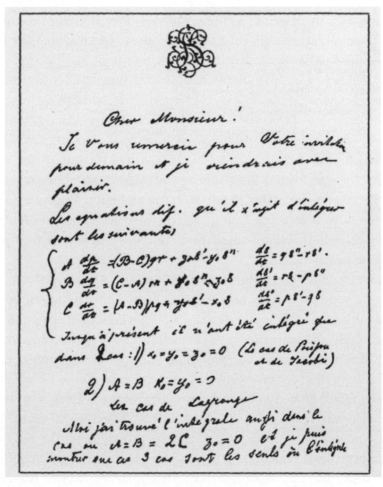

A letter from Sophie Kowalevski, reporting her results.

variables to describe how such an object moves. This is where theta functions came in. Their main feature is that they can combine several variables together, which makes it seem that there are fewer changing parts. Using theta functions, Kowalevski was able to reduce the two variables to one and then apply some of the techniques that Euler and Lagrange had developed. As soon as she recognized the applicability of theta functions to the mathematical mermaid, she knew she was on the right path. She wrote a letter sharing the result and her desire to make further progress.

The news soon spread. Members of the French Academy of Sciences were thrilled to hear of the breakthrough and decided to set the problem for the selection of the prestigious Prix Bordin that year in the hope of enticing her to submit her work. In 1886, the task was announced pithily:

PRIX BORDIN
(Question to be solved by the year 1888)
'Improve, in some important point, the theory of the movement of a rigid body.'

Up for grabs was a huge amount of mathematical prestige and 3,000 francs (more than £13,000 in today's money). Around a dozen people entered the competition, each submission anonymous. A crack team of well-established mathematicians was selected to form the judging panel.

The deadline was 1 June 1888. However, Kowalevski suffered another bereavement: her beloved sister Anna died at the age of forty-four in October 1887 and, grieving, Kowalevski missed the deadline. Fortunately, the committee at the French Academy were accommodating. Although the prize was presented as an open competition, they had chosen this particular problem with Kowalevski's work specifically in mind. Nobody else had even got close to solving it to their knowledge, and so they were willing to wait. When she submitted her half-complete manuscript and requested an extension, she received the reply: 'Since these Academicians [the evaluators of the Prix Bordin] take vacations, of which they have great need, you may rest assured that they will not set it before the month of October.' She was granted a three-month extension and managed to submit her work within that time.

When the evaluators gathered that autumn, the judges were so impressed they not only awarded her entry the prize but upped the prize money to 5,000 francs (roughly equivalent to £22,000 today). Her paper had her motto written on both the envelope and the paper itself. 'Say what you know, do what you must, and whatever will be, will be.'

Grand prix

At the recommendation of Chebyshev, Sophie Kowalevski became the first female member of the Russian Academy of Sciences – although she was made only an associate member, unlike her male colleagues, who were full members.

Kovalevskaya died from pneumonia in 1891, at the age of just forty-one. She had faced many challenges in her life, and had overcome them, and there were challenges also to her legacy. The influential mathematician Felix Klein had consistently disregarded and undermined her work while she was alive, claiming that all she was doing was mimicking the mathematics of her supervisor, Karl Weierstrass. In 1926, he repeated these claims in his history of nineteenth-century mathematics, leaving the impression that her work was not original: 'Her works are done in the style of Weierstrass and so one doesn't know how much of her own ideas are in them'.[13] Klein believed her primary contribution lay in making Weierstrass more confident in his own ideas.

Nearly eighty years later, the prolific mathematician and writer Steven Krantz, in *Stories and Anecdotes of Mathematicians and the Mathematical*, repeated hearsay that she had an affair with her married colleague, Gösta Mittag-Leffler, despite the evidence being extremely thin, writing, 'she lived in Mittag-Leffler's house for some time, and it was rumoured that they were intimate (she was *not* his wife)'. He goes on to put emphasis on her physical appearance, describing her as an 'exceptionally beautiful woman' whose 'physical attraction no doubt contributed to her appeal'. The book includes an image of an

unknown woman in fancy dress, captioned 'the lovely Sonja Kow-
alewska dressed up as a kitty kat'.[14]

Others have also placed undue emphasis on Kowalevski's looks. In a
letter to Kowalevski, Weierstrass mentions that Kowalevski had convinced German chemist Robert Bunsen to change his view that no
woman should be allowed in his laboratory, although no specific details
are given as to how.[15] Some historians have suggested that it was her
beauty that altered Bunsen's opinion, instead of, say, her reasoning
skills and intelligence. E. T. Bell, a twentieth-century mathematician
and writer, unexpectedly included her in his *Men of Mathematics*,
describing her as a 'dazzling young woman'[16] perpetuating the por-
trayal of her as someone whose appearance was more important than
her mathematics. As historian Eva Kaufholz-Soldat writes, the representation of Kowalevski as some sort of femme fatale has been repeated
by many of her biographers.[17]

Some writers have also hinted at a romantic 'relationship' between
Kowalevski and Weierstrass. It's true that Weierstrass had romantic
feelings for her, but there is no convincing evidence that this was
reciprocated. These portrayals add up to a general impression that
some of Kowalevski's success was down to her ability to use her looks
to manipulate other great mathematicians and scientists, rather than
simply having the mathematical talent herself.

There is currently little evidence that Kowalevski was romantically involved with Mittag-Leffler or Weierstrass, but even if that
changed it shouldn't alter our view of her as a mathematician in her
own right. Mathematics has never and will never be removed from
personal relationships, just like any other occupation. It is easier to
publicly praise the work of people you like than the people you don't.
Or to tell a friend about a vacant role or to nominate a collaborator
for membership of a professional society. This is understandably
commonplace and also what makes it all the more difficult for an
outsider, such as Kowalevski, to break in.

Kowalevski succeeded in a system that was not set up for her. But
we shouldn't reduce her story to one only of struggle. From the
first moments that she laid eyes on the mathematical equations
pasted on to the walls of her childhood, she had a deep fascination

with mathematics and its ability to uncover hidden truths, and this drove her more than anything else. As she wrote: 'It seems to me that the poet has only to perceive that which others do not perceive, to look deeper than others look. And the mathematician must do the same thing.'[18]

This idea of looking deeper than anybody had looked before is an important one. Well-established mathematical ideas can sometimes portray a veneer of completeness, as if there is nothing more to say. But when an intrepid mathematician or two is willing to look a little closer, sometimes there is more to see than could have once been imagined. And this is exactly what happened when the 'Copernicus of geometry', among others, re-examined the fundamental rules of shapes Euclid and others had laid down thousands of years before. It all began with a tale about Gauss.

12. Revolutions

There is a story about German mathematician Carl Friedrich Gauss that goes something like this. In 1821, he was on a mountain, setting up his newly invented heliotrope – an instrument for reflecting and focusing sunlight over large distances. The instrument gave surveyors precise points to reference when calculating the angles between different things. In this case, the things were the peaks of three mountains: Hohen Hagen near Göttingen, Brocken in the Harz Mountains and Inselsberg in the Thuringer forest (see map below). The resulting triangle was massive, with its longest side around 110 kilometres long – exactly as he had planned. These three mountain peaks were going to help Gauss question the very nature of triangles.

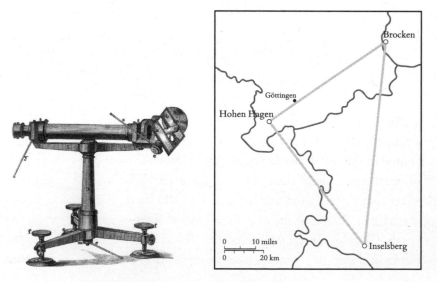

Gauss's heliotrope (*c*.1822) (*left*).
Gauss's great triangle (*right*).

Gauss knew that Earth was curved★ and as such the angles in the triangle should not add up to 180 degrees but to more than this. Even at this mountainous scale, the difference would be minuscule, but as he wrote to a colleague, Olbers, on 1 March 1827, 'the honour of science demands that one understand the nature of this unevenness clearly'.[1]

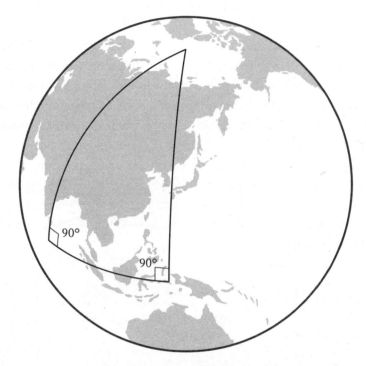

An example of a triangle on Earth.

'Geometry' literally means 'measurement of the Earth'. Yet in Europe in the early nineteenth century, geometry was still strongly rooted in Euclid's *Elements* and the postulates he wrote down thousands of years ago. One of them, known as the parallel postulate, essentially states that the angles in a triangle always add up to 180 degrees, which is true in a flat plane but not on a curved surface such as that of the Earth. Clearly, something was amiss. What followed would be a slow but radical shift in geometry and its implications, a complete rewriting

★ 'An oblate spheroid!' shouts Newton from the audience.

of which geometries, shapes and spaces were mathematically possible. These changes would lay the foundations for the biggest upheaval in the physics of the universe since Newton.

Euclid goes viral

For more than a millennium, Euclid's *Elements* was the most important mathematical text in Europe. Since the thirteen-book treatise was first produced, around 300 BCE, it has ranked second only to the Bible in terms of the number of editions published. It has been translated repeatedly, and there were even pocket-sized and pop-up versions.

 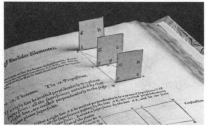

Billingsley's version of Euclid's *Elements* (1570).

The book begins with the five postulates which outline the basic assumptions of geometric concepts such as points, lines and distance that we touched on in Chapter 1. The first four postulates were simple enough, for example, 'a straight line may be drawn from any one point to any other point', but there was something about the fifth – the parallel postulate – that had irked mathematicians for millennia. It states that if two lines are drawn which intersect a third in such a way that the sum of the inner angles on one side is less than two right angles, then the two lines must inevitably intersect each other on that side if extended far enough. It's a little hard to imagine, so perhaps it is better depicted. The two lines with arrows in the diagram overleaf, the postulate states, will eventually meet.

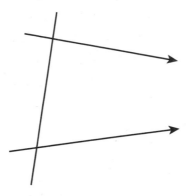

Another way of putting this is that if the two lines both intersect the third so that the sum is 180 degrees, i.e., they are parallel, they will never meet.

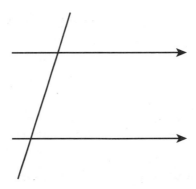

That may seem intuitive, but this fifth postulate was more complicated than the other four. For years, many mathematicians, both in the Middle East and in Europe, including Ḥasan Ibn al-Haytham (*c.* 965–*c.*1040), Omar Khayyam (1048–1131), Giovanni Girolamo Saccheri (1667–1733) and Johann Heinrich Lambert (1728–77), had hoped there would be a way to deduce it from the first four postulates, but their efforts were in vain. Unusually, this was a case where the best approach seemed to be simply to give up.

One person who did the right thing and gave up was János Bolyai. But his story is not a straightforward tale. Bolyai is part of a trio of mathematicians who hit upon similar ideas at similar times. The exact relationship between them is fuzzy – like a triangle with angles that don't add up to 180 degrees – and is an important illustration of the ways in which mathematical ideas often develop.

Born in the Hungarian town of Kolozsvár (now in Romania) in 1802, János showed mathematical promise in his younger years. His father, Farkas Bolyai, just happened to be best friends with a colleague of Gauss, and asked him if his son could come to live with him to learn mathematics. It was an unusual request, and Gauss declined. Instead, with few options available to study mathematics, János Bolyai went to study military engineering at the Academy of Engineering in Vienna at the age of sixteen. He finished the seven-year course in four years and joined the military.

In his spare time, he began working on replacing the parallel postulate with something else. His father had worked on this before him, but it had turned out to be the bane of his existence. 'For God's sake, I beseech you, give it up,' his father wrote to him. 'Fear it no less than the sensual passions because it too may take all your time and deprive you of your health, peace of mind and happiness in life.'² János eventually did give up, but not in the way his father had expected. He gave up on the parallel postulate as a building block but started to build a version of geometry without it. In 1823, he announced to his father that he had 'created a new, different world out of nothing'.³ János Bolyai had realized that if the parallel postulate does not hold, there could be a completely different type of geometry. He called it 'absolute geometry'.

Imagine the surface of a saddle or a stackable potato-based crisp.★ As a straight line is defined as the shortest distance between two points, on such a surface straight lines appear curved and triangles are squashed, which means that the sum of the three angles is less than 180 degrees. Imagining what happens on surfaces like this would later become known as hyperbolic geometry. It may, to us, seem like just a bit of imaginative fun, but it would have been very difficult for people to understand at the time. Euclid's postulates and the geometry they produced were sacrosanct. Anything that suggested that

---

★ You know the one . . . comes in a tall cylindrical tin.

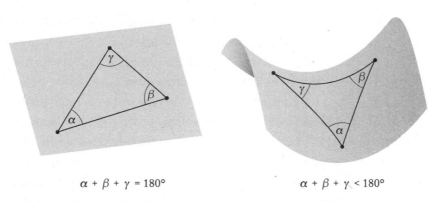

$$\alpha + \beta + \gamma = 180°$$ $$\alpha + \beta + \gamma < 180°$$

Euclidean geometry (*left*) and an example of absolute or hyperbolic geometry (*right*).

other geometries could exist was a challenge to the Euclidean orthodoxy. Perhaps, as a result, János's father was distinctly unimpressed.

A few years later, János was posted to Arad, Romania, and met his
former mathematics teacher from the Academy of Engineering in
Vienna, Wolter von Eckwehr, there. He passed von Eckwehr a manuscript on his ideas, but von Eckwehr failed both to provide any
comments, and to even return it. János's father had become more
accustomed to his son's ideas in the seven years since he first saw them
and suggested that the work should be published as an appendix to his
forthcoming book. Farkas also sent a report about his son's ideas to
his old friend Gauss. However, far from being the expected moment
of triumph, Gauss replied coldly that the entire content of the work
coincides 'almost entirely with my mediations, which have occupied
my mind partly for the last thirty or thirty-five years'.[4] Back in 1824,
Gauss had written to a colleague stating that he saw a 'curious geometry, quite different from ours, but thoroughly consistent'.[5] Gauss
had indeed been working on the same problem.

János was devastated to hear that Gauss had known of something
similar, and the incident set off a lifelong dislike of the man. Yet, despite Gauss's confidence, he had not been the first to discover this new
geometry either. There must have been something in the air at the
time, subtly encouraging mathematicians to challenge the old geometric ideas. Nikolai Lobachevsky had beaten them both to it.

When parallel lines meet

Nikolai Lobachevsky first studied mathematics in Kazan, Russia, after he and his brother won government scholarships to attend Kazan Gymnasium, an elite high school. After five years, he entered Kazan University, which was filled with professors who had been educated in Germany. Despite Russia being a hotspot for mathematicians at the time, few of them had been born in Russia. One of the professors, Martin Bartels, taught courses on mathematics and its history, and it is probably in one of these classes that Lobachevsky first came across the parallel postulate.

Lobachevsky graduated from Kazan University with master's degrees in physics and mathematics, and began to teach there. He directed the construction of a new astronomical observatory, which became one of the best equipped in Russia. By 1816, he had turned his attention to Euclid's axioms and decided to try to use the world around him to understand them better. Only the geometry of the universe, he thought, would be able to tell him about the fundamentals of geometry.

Lobachevsky decided to look at three stars, two of which were relatively nearby and one that was much further away, and attempt to calculate the sum of the angles between them. Dealing with such a great distance, however, was not easy – even at one of the best observatories in Russia. He did manage to take measurements demonstrating that the sum of the angles in the triangle of stars was less than 180 degrees, but the difference was extremely small – so small that he dropped the measurements from a working manuscript on the underpinnings of geometry and instead went with a theoretical argument. This argument led him to the same uncomfortable position Bolyai would later encounter, where Euclid's parallel postulate does not hold. Unable to prove it with data, Lobachevsky called this mathematical world 'imaginary' geometry.

He first submitted his work to the St Petersburg Academy of Sciences, but it was rejected, which was no surprise – in many ways, the work was scandalous. A world in which the parallel postulate does not hold might look more like an Escher artwork than our perception of reality. In Euclidean geometry, given a line and a point, there is a single parallel line that can be drawn through that point, but in hyperbolic geometry there are infinitely many. This was as shocking to people then as the idea that the sun does not go around the Earth was to people in the 1500s.

Lobachevsky was undeterred by this rejection and began to call his work pangeometry, positing it as a general geometric theory which included Euclid's geometry as a particular case. He published the work in his university's journal, *The Kazan Messenger*, in 1829. Because the ideas were so revolutionary, it was decades until they were accepted in Russia. However, as we saw with Bolyai and Gauss, there were others reaching similar conclusions. It was Lobachevsky, though, who would later become known in mathematical circles as the Copernicus of geometry.

Bolyai, Gauss and Lobachevsky had hit upon only one side of the story. In their work, they had explored hyperbolic geometry on a saddle-shaped surface. We live on the surface of a sphere, where

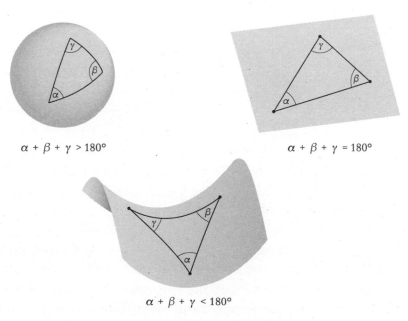

$\alpha + \beta + \gamma > 180°$ $\alpha + \beta + \gamma = 180°$

$\alpha + \beta + \gamma < 180°$

(*Top left to bottom*) positive curvature, zero curvature and negative curvature.

geometry behaves not only in a 'non-Euclidean' way but also with a different sort of curvature than the one the trio had hit upon. It would be Bernhard Riemann who would turn his attention to this form of geometry, known as positive curvature.

In 1854, Gauss was supervising Riemann's postdoctoral qualification at the University of Göttingen. Riemann explained that there was a geometry that worked in a way opposite to Bolyai's and Lobachevsky's 'hyperbolic' geometry. Instead of the angles of a triangle adding up to less than 180 degrees, they added up to more. Gauss was aware of this sort of geometry, but Riemann was the first to really explore its mathematical implications.

Elliptic geometry violates not just the parallel postulate but three of Euclid's other postulates. To recap: the first postulate says that, given two points, there is a unique straight line that connects them. But imagine drawing a line that follows the shortest distance between the north and south poles, i.e. a straight line. In Euclidean geometry, there should be only one option, but, in fact, any of the lines of longitude would work. It's not just that there is no longer a unique line connecting two points – in some cases in elliptic geometry, there are infinitely many!

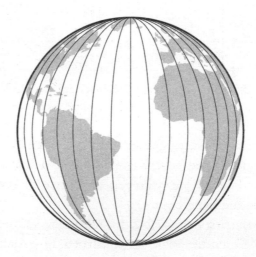

Furthermore, Riemann noticed that many ideas in geometry can be generalized to higher (and lower) dimensions. Take the ideas of a square and a cube. If you place a square at a right angle to every side of another square, you end up with a cube, but why stop there? If you place a cube at every face of a cube and connect them together, you end up with another shape — a tesseract. You need four dimensions to create this, but mathematically speaking, that's no problem. This led mathematicians to a more generalized idea of a square, one that can apply in higher dimensions.

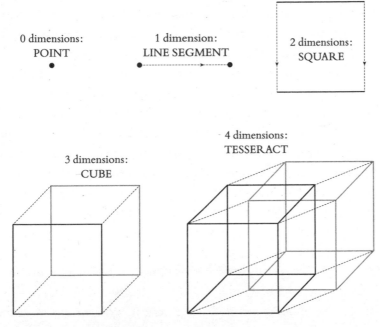

0 dimensions:
POINT

1 dimension:
LINE SEGMENT

2 dimensions:
SQUARE

3 dimensions:
CUBE

4 dimensions:
TESSERACT

'Squares' in different dimensions. Mathematicians sometimes call these (*top left to bottom right*): 0-cube, 1-cube, 2-cube, 3-cube and 4-cube.

The University of Göttingen would soon become a hub for non-Euclidean geometry, and Felix Klein would be one of the leaders of the movement. It was Klein who classified the different new geometries as elliptic and hyperbolic.

He was also the first to describe a strange higher-dimensional shape known as a Klein bottle, which is constructed by sticking together the two ends of a cylinder in 'reverse orientation'. By giving the cylinder a twist when you reconnect it, you can create a shape that has only one side. This cannot be constructed in three dimensions

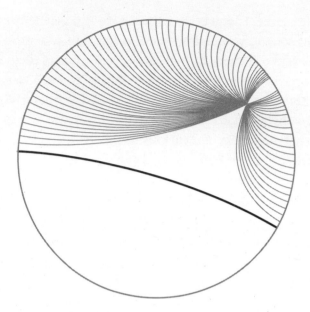

There are infinitely many parallel lines through a
single point in hyperbolic geometry.

but, if you have four dimensions, it can. The Klein bottle is a higher-
dimensional version of the perhaps more familiar Möbius strip. The
surface of a Möbius strip is created by twisting a rectangle and sticking two of the edges together. The result gives a flat surface with
only one side. If you could drive along it, you would eventually
cover the whole surface without ever having to pick up your car and
jump over to the other side.

David Hilbert, another mathematician at Göttingen, in 1899 proposed a new set of postulates to replace Euclid's. Among Hilbert's
twenty new axioms was one about parallels. Rather than universally
describing parallel lines, as Euclid had attempted, Hilbert's version of
the parallel postulate began, 'In a plane . . .' In other words, parallel
lines were assumed to exist on a two-dimensional plane, but in higher
dimensions and on other surfaces there was no reason to assume that
they existed. Hilbert ushered in a new era of mathematics, topped off
by a speech made in 1900 at the recently created International Congress of Mathematicians (ICM). He gathered twenty-three open

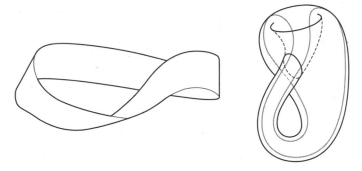

Möbius strip (*left*) and Klein bottle (*right*).

problems, those he considered to be the most important to resolve, and announced ten of them at the ICM. His speech helped define the direction of mathematical research for decades to come.

A fundamental speed limit

By 1900, the world had gone through major upheaval, with industrialization, the Atlantic slave trade and the Opium Wars between Qing China, the UK and France. The wars exacerbated a period of decline in China and a subsequent opening of the trade ports of Shanghai and Hong Kong. Japan was also going through rapid change, as the country was forced to end its seclusion policy in 1853. For over 250 years, only very limited numbers of Dutch and Chinese people had been allowed to visit Japan for trade (people from elsewhere were barred entry), and very few Japanese people were permitted to travel beyond the country. The US navy had sent four warships to Japan's shores and demanded that it open up trade to the West. The governing regime in Japan was taken aback by the sheer military strength on display, which kickstarted a massive debate about Western ways, and whether they should be adopted.

Euclid's geometry had already made its way to Japan through Chinese translations of the *Elements*. However, up until that point there were no academic institutions specializing in mathematics, so few people understood it. It became official government policy to prioritize learning *yosan* (Western mathematics) over *wasan* (Japanese mathematics). Towards the end of the nineteenth and the beginning of the twentieth century, the Japanese government sent scholars to the US, the UK, France and Germany. They came back and formed scholarly associations, taught at universities and wrote textbooks. Western mathematics became the norm. So much so that, beyond the abacus, very little traditional Japanese mathematics was still in use. Japan then became an exporter of Western mathematics to the areas of Asia it had once colonized – parts of which are today in Taiwan, Micronesia, North and South Korea and China.

Mathematics was developing rapidly – and physics would shortly have its turn. In 1905, Albert Einstein published his special theory of relativity, completely rewriting our understanding of time and space. Up until then, speed had been viewed as relative. If you throw a ball from a moving car, the ball moves at the speed you have thrown it plus the speed of the car, however earlier experiments and explorations had provided hints that light *didn't* behave this way – its speed in a vacuum is always constant, no matter what.

This observation had stumped many physicists, but Einstein's light-bulb moment was to stop thinking of it as an inconvenience and instead think of it as a fundamental speed limit for the universe. Writing out the equations led him to incredible results; for example, he showed how energy and matter could be exchanged through the relationship of $E = mc^2$, where E is energy, m is mass and c is the speed of light in a vacuum. It also led him to the strange conclusion that spacetime was not a single homogeneous entity but instead had to warp to accommodate this universal speed limit.

This certainly brought him into the realms of non-standard geometries, but it was when he added gravity to his theory that he really found himself immersed in it – his theories would simply not hold otherwise. In his general theory of relativity, Einstein stated that gravity is the result of massive objects bending spacetime, and these bends mean that objects in spacetime provide acceleratory forces on each other linked to their mass. Between this book and you, this force is minuscule. But between Earth and you, this force is what keeps you on the ground. Underpinning Einstein's mathematical workings was the geometry that Riemann had developed. Einstein had made the conceptual leaps required for his theories of relativity, but it would have been a lot harder if the mathematics hadn't already been worked out. Ultimately, his approach was all about axioms. Taking the principle that light has a speed limit to its logical conclusion led him to special relativity and beyond. But physics and mathematics had many other axioms too. And out of the process of working out which were really basic enough to be axioms and which could be deduced came one of the most awe-inspiring theorems of all time.

The equations of the universe

Emmy Noether was born in 1882 in the Bavarian town of Erlangen, where her father was a mathematician at the local university. She had originally thought she might like to be a French and English teacher but soon switched to mathematics. After completing a doctoral degree, she got a job at the Mathematical Institute of Erlangen. Like for Sophie Kowalevski early in her career, Noether's 'job' was unpaid. Her family had supported her financially throughout her studies and continued to do so during her work at the institute. Her big break came when David Hilbert and Felix Klein, both working at the University of Göttingen, paid her to assist them on problems arising from Einstein's general theory of relativity. And when she arrived there, she presented them with what is now known as Noether's theorem.

To see just how stunning the theorem is, imagine writing down all the equations that describe the universe – a truly baffling number, involving a seemingly improbable number of scientific disciplines. And regardless of what they are, Noether's theorem tells us something about all of them.

Her theorem says that if the laws of physics apply equally everywhere in the universe, then momentum must always be conserved. And it says that if the laws of physics today are the same as those tomorrow, then energy must be conserved too. Even if we don't know exactly what the equations of the universe are, Noether's theorem tells us about their consequences. Before Noether, the conservation of momentum and energy were assumptions. Now they could be deduced.

Noether's proof of her theorem was a tour de force built using tools from a subject known as the calculus of variations. Think of it as a generalization of calculus that doesn't just apply to mathematical functions that act on numbers, but to mathematical functions that act on functions themselves. Noether was able to use the approach to mathematically capture the essence of physical laws and symmetries without ever having to get into the nitty-gritty of what they were. She could then manipulate these expressions to prove her theorem, revealing the fundamental link between them.

This was an incredible achievement, but it would not lead her down an easy path to a career in mathematics. After gaining her doctorate at the University of Erlangen, Hilbert was keen to get her a position at the University of Göttingen, which boasted many leading mathematicians and was then the growing epicentre of mathematics in Europe, thanks to the presence of Hilbert and Klein. But her proposed appointment was met with protests by other academics, who thought there could be no place in academia for women. Exasperated, Hilbert said to the university higher-ups considering the application, 'I do not see that the sex of the candidate is an argument against her . . . After all, we are a university, not a bathhouse!'[6] Eventually, Hilbert did manage to get Noether a position, as a sort of guest lecturer – another unpaid position. At the university-not-bathhouse, many of the mathematicians enjoyed swimming and talking in the men-only pool. Noether was not allowed to join them, but allegedly took up swimming there anyway.

Four years later, in 1923, Noether became a stipendiary lecturer, a slight improvement, as the position came with a small payment. However, by 1933 the Nazis were in power in Germany and rumours started to spread that Noether, who was Jewish, would be dismissed from the role. Many mathematicians and graduate students wrote letters in protest. They testified to Noether's superb mathematical research abilities and commended her commitment to teaching. One of Noether's collaborators, Helmut Hasse, wrote, 'I am convinced that Miss Noether is one of the leading mathematicians in Germany.

Especially for the young generation, it would be a very heavy loss if Miss Noether is forced to move abroad.'[7] By 31 July, letters had come in from all over the world, including Copenhagen, Vienna, Cambridge, Bologna, Zurich, Osaka and Tokyo. A few months later, Noether received confirmation from the Prussian Ministry for Science, Art and Popular Education that she would be dismissed. However, she had already planned her escape.

Mathematicians in Nikolausberg near Göttingen in July 1933.
(*From left to right*) Ernst Witt, Paul Bernays, Helene Weyl,
Hermann Weyl, Joachim Weyl, Emil Artin, Emmy Noether,
Ernst Knauf, unidentified person, Chiungtze Tsen and Erna Bannow.

Noether had also received offers to move to both Oxford and Moscow, but she ultimately took up an offer from Bryn Mawr in Pennsylvania – the only women's college at the time with a doctoral programme in mathematics. There, she found herself with several female colleagues for the first time. She became good friends with Anna Pell Wheeler, the head of the mathematics department, who had played a large role in bringing her to the US. Wheeler knew how hard it could be for women in mathematics. 'There is such an objection to women that they prefer a man even if he is inferior both in training and research,' she wrote.[8] She had previously studied at Göttingen and so spoke German, which may have helped increase the appeal.

The pair became friends and collaborators. Noether's years at Bryn Mawr were some of her happiest; she felt more appreciated professionally there than she ever had at home. With the support of Wheeler and others, she threw herself into teaching. Her English was rudimentary, but she made it through, occasionally switching to German if her initial attempts at explanation didn't work. She was invited to give guest lectures at the Institute for Advanced Study (IAS) at Princeton, a new research institute created with the aim of bringing together the greatest minds of the day. Though it was men-only, they made an exception for Noether, inviting her (although only as a 'visitor', rather than as a 'professor') to give weekly lectures in the school of mathematics.

When she was in Germany, Noether's expertise had gained her a group of followers who became known as Noether's boys. Some had come from far afield to learn from her. Among them was Chinese student Chiungtze C. Tsen, who had come to Göttingen from Qing China after winning a scholarship to study in Europe. He would go on to prove a basic result about functions involving curves that came to be known as Tsen's theorem. Also among them was Teiji Takagi from Japan, who studied mainly with Hilbert as part of the country's push to better understand Western mathematics but was one of those who wrote letters of protest when it seemed that Noether would lose her position. Noether loved teaching, often following no strict lesson plans and preferring open conversations. On one occasion when the university was closed taking these discussions to a coffee house. At Bryn Mawr, she acquired a similarly devoted set of followers. However, this time, Noether's followers were young women, among them Olga Taussky, who would go on to publish over three hundred mathematical research papers.

Noether would spend only a couple of years at Bryn Mawr. She died in 1935, aged fifty-three, days after undergoing surgery to remove a tumour. She had told only her closest friends that she was ill and so her death came as a shock to many. As the news became public, tributes came pouring in, including one from Albert Einstein, who wrote in the *New York Times*: 'Fräulein Noether was the most significant creative mathematical genius thus far produced since the higher education of women began.'[9]

By the time Noether had left Germany, she was a famous and renowned mathematician. Though it is unclear if she knew very much about Bryn Mawr before she joined, everyone there certainly knew a lot about her. Bryn Mawr had been keen to help her and was able to do so by means of a programme run by the US Institute of International Education to help displaced German scholars. Many other displaced scholars were invited to various institutions. However, plenty of places would not accept people of Jewish descent. Others refused to support anyone from Germany at all. This reluctance was a reflection of societies at large that were far from open and inclusive. Noether had done a great deal to change views on women in mathematics, greatly expanding what people believed was possible. However, diversity in mathematics still had a long way to go. A great expansion of who got the opportunity to study and pursue the subject would eventually happen, though it would be hard won.

When budding astronomer Benjamin Bannaker got wind of what was said about his predictions and calculations by the established astronomer David Rittenhouse, he was understandably a bit irritated.

It was the early 1790s and Bannaker was attempting to publish an almanac that included timings for eclipses and when planets would align in the sky. Andrew Ellicott, who was famous for mapping parts of the area west of the Appalachian Mountains in the newly named United States of America, was supportive of the endeavour. He had arranged for others to take a look in the hope of getting endorsements for Bannaker. One such endorsement came from Rittenhouse, who we briefly met in Chapter 9 as an advocate of Newtonianism. He was already established in the field of astronomy, having written his own almanacs, and was clearly impressed with Bannaker's work. He described the calculations as 'sufficiently accurate'. He also said that it was a 'very extraordinary performance'. But he tempered his view of just how extraordinary the performance was with the following words: 'considering the colour of the author'.[1]

Bannaker was born in Maryland. His mother, Mary, was a free Black woman and his father, Robert, a former enslaved person from Guinea. There are conflicting accounts of Benjamin's broader ancestry, but one suggestion is that his maternal grandfather may have been a member of the Dogon people from West Africa. The Dogon's religion features details about the solar system that are impossible to discern with the naked eye, such as the rings of Saturn and the moons of Jupiter. It's not known exactly how the Dogon made these discoveries, and Bannaker's grandfather died before he was born, but it's possible he passed on this knowledge to his wife (Bannaker's grandmother), who then passed it to him, kickstarting a fascination with astronomy.

The early details of Bannaker's life are a little hazy. It's thought that as a young boy he went to a school and learned reading, writing and arithmetic, but that he probably stopped attending when he was old enough to work on the family farm. At the age of twenty-one, he managed to completely re-create the mechanics of a borrowed pocket watch by carving to-scale versions of the components out of wood. The wooden clock chimed on the hour and continued to work until his death, over fifty years later. The clock was particularly remarkable considering that so few people in the Americas kept track of time on an hourly basis. David Rittenhouse and several other Americans had created elegant clocks for wealthy families, but they were still a rarity. Bannaker was also fascinated by the way the planets and the stars moved and purchased a small telescope to study them more closely.

From his late twenties, Bannaker lost much of his passion for astronomy, perhaps due to a romantic episode with a tragic ending. He worked on a farm until, in his late fifties, he met the Ellicott family. They were Quakers and believed in racial equality, and gave him a job at a mill they owned. While there, he borrowed their astronomy books and quickly mastered the algebra, geometry, logarithms and trigonometry contained in them – so much so that only a year later he was producing his own predictions of solar eclipses. The family recognized his mathematical skill and Andrew Ellicott asked him to help with surveying the land. Bannaker produced astronomical calculations for the surveys, as well as calculating various points that could be used to set federal district boundaries. By 1791, Bannaker had relinquished this work and set about producing his astronomical almanac for the following year. However, initially, no publisher would accept the book, and Andrew Ellicott stepped in to help.

Rittenhouse, who was also against slavery, had clearly endorsed Bannaker's work. One reason he had for bringing race into his assessment was that he wanted to counteract the prevailing view that Black people were of inferior intelligence to white people. However, Bannaker wanted his work to be assessed on merit alone – and Rittenhouse's backhanded compliment bothered him. He is said to have responded, 'I am annoyed to find that the subject of my race is so much stressed. The work is either correct or it is not. In this case, I believe it to be perfect.'[2]

Bannaker's work was indeed correct, and with the help of endorsements from Rittenhouse and others he landed a deal to publish astronomical almanacs for the next six years. The project was a commercial success, with as many as twenty-eight editions being published in five different states.

Benjamin Bannaker's portrait featured on the cover of an edition of his 1795 almanac.

In the 1800s in the US, few Black people had the opportunity to study and pursue mathematics. Many private colleges had no explicit rules forbidding Black people, but they very rarely admitted any, and although several institutions provided Black clerics with a Christian education, they didn't provide an opportunity for further academic training. In 1857, the all-white and all-male Supreme Court declared that Black people were not and could never become American citizens, and this fuelled the fire that led to the American Civil War and the subsequent, and ongoing, struggle for emancipation.

After this period, the US developed into both a political and a mathematical powerhouse. American universities became world-renowned as research hubs for new mathematics, particularly the mathematics required for the information age. Yet many institutions continued to be near-impenetrable fortresses for Black mathematicians. Breaking down the barriers would not be easy.

Civil rights

In the 1870s, Jim Crow laws created an educational system that segregated 'coloured' people from white people, providing a legal basis for discrimination by race. These laws were challenged by groups such as the National Association for the Advancement of Colored People (NAACP) and activists like Oliver Brown and Linda Carol Brown but ultimately stayed in place for around a hundred years. Higher educational institutions now known as HBCUs (historically Black colleges and universities) were set up to serve the needs of Black Americans and would eventually number over a hundred, enabling thousands of people to access an education that would otherwise have been unavailable. The HBCUs were far from perfect, having few Black people in leadership and management positions, but they provided a particularly important service until the US Civil Rights Act of 1964 outlawed discrimination on the grounds of race, colour, religion, sex and national origin.

One mathematician who would thrive at a HBCU was Elbert Frank Cox. Born in 1895 and originally from Indiana, Cox was a talented violinist who also showed acumen in mathematics and physics. He followed in his father's footsteps by entering the predominantly white Indiana University in 1913 to study mathematics. This was rare, but not completely unheard of. He finished his degree, obtaining straight As in all his papers, although these results were obscured on his transcript: the word 'COLORED' was printed across it – the norm for Black graduates at the time.

Cox went to France to serve in the US army during the First World War. On returning to the US, he taught mathematics in public schools in Kentucky and North Carolina. Cox was an outstanding teacher and was given a fellowship which enabled him to enter Cornell University to study mathematics at doctoral level.

Cornell's founder, Ezra Cornell, was one of the early opponents of slavery, and so the university accepted a few African American students each year. At Cornell, Cox met his thesis adviser, William Lloyd Garrison Williams, who was leaving for McGill University in Canada. Williams noticed Cox's ability, and Cox visited him in Montreal, finalizing his doctoral thesis there and then submitting it back in the US. In his thesis he worked on the solutions to a type of equation linked to Bernoulli numbers. These numbers, which turn up in many disparate areas of mathematics, take their name from the seventeenth-century Swiss mathematician Jacob Bernoulli but were independently discovered around the same time by Japanese mathematician Seki Takakazu.

Cox's thesis was approved in 1925; he was the first Black person in the world to receive a Ph.D. in mathematics. Though it was a landmark moment, Cox still struggled to get his work published. Williams urged Cox to send his thesis to a university outside of the US for publication, but universities in both the UK and Germany turned him down, purely because of his race. Eventually, he sent it to Japan, where Tohoku Imperial University recognized its quality and published it in the *Tohoku Mathematical Journal* in 1934.

After a brief stint at West Virginia State College, Cox was hired in 1930 by the HBCU Howard University in Washington DC. He

A photo of the members of the 'Euclidean Circle' in 1916. The group was put together for people in the mathematics department at Indiana University to discuss and share ideas. Cox (*front row, far left*) was the first African American student to join.

largely taught and supported students during this period, then went on to teach an engineering science and army training programme in the latter years of the Second World War. He chaired Howard's mathematics department from 1957 to 1961, and published two research papers in the course of his career, one on particular equations and their solutions and the other a mathematical analysis of three different ways to grade students. Most of his time was spent teaching the next generation of mathematicians, and he certainly seemed to have a talent for it, however as a professor at a HBCU he would have had very little choice in the matter. HBCU professors tended to have far higher teaching workloads, and inadequate financial support meant that finding time to pursue research was near impossible.

Although HBCUs did contribute to higher educational levels among Black people in the US, research positions and doctoral studies were still largely blocked. So Cox worked with one of his colleagues, Dudley Weldon Woodard, the second African American to earn a Ph.D. in mathematics, to establish a master's programme at Howard. Cox was said to have overseen more master's students than anyone else at the university, and the pair built up a group of talented young mathematicians who were ready to pursue doctoral research. All they needed was a doctoral programme. Cox, sadly, did not live to see his dream realized, but he was integral to making it possible. As a 1969 obituary in the *Washington Post* put it: 'Cox helped to build up the department to the point that the PhD program became a practical next step.'[3] The combination of his prestige as a research mathematician and the people he attracted to Howard made the establishment of a Ph.D. programme almost inevitable, and in 1976, Howard's Ph.D. programme was finally set up – the first mathematics Ph.D. programme at any HBCU.

Education for all

Washington DC, where Howard University is located, was also home to one of the first Black women to receive a Ph.D. in mathematics, Euphemia Lofton Haynes. However, her career did not have mathematical research or one particular institution as its focus, but racial equality. She played a crucial role in the wider movement across the United States, trying to redress the systemic racism that was preventing equal access to education.

Haynes was born in 1890 into a wealthy family who were part of the so-called 'Colored 400', community leaders in Washington DC who helped unite people of colour during segregation. She attended one of the first high schools for African American students in the US. Established in 1870, many of the graduates of M Street High School had already become Black leaders in DC and beyond. As valedictorian in 1907, she delivered a speech that included the lines, 'For a person of intelligence is well equipped to solve the problems of life . . . We must have some defined aim in life and be able to fill competently that position in which we may find ourselves.'[4] After graduation, she attended Smith College to study mathematics and psychology, then moved to the University of Chicago, which had a slightly more liberal outlook than other white universities at the time. Between 1870 and 1940, forty-five African Americans earned doctoral degrees there. This was still a small number, but it was more than anywhere else in the country.

Haynes completed her master's in Chicago then became a professor of mathematics at Miner Teachers College. Alongside this she took graduate-level mathematics classes at the University of Chicago and started work on a Ph.D. thesis in the field of algebraic geometry. However, this thesis was never intended to catapult her into a career in mathematical research. Her supervisor, Aubrey Edward Landry, often set students open problems that were not mainstream research but suitable for those who went into Ph.D. programmes hoping to teach afterwards. After graduation, she chaired the department at Miner Teachers College and also took on a part-time job as a teacher at Howard University.

In 1954, the US Supreme Court ruling *Brown v. Board* decreed that unequal schooling rights must be eliminated, and so started the long process of desegregation. A four-track curriculum was introduced: two intended to prepare students for college; one for blue-collar work; and one for those deemed to be academically behind their peers. IQ test scores or the opinion of a teacher were used to assign

Euphemia Lofton Haynes.

students to a particular track and, once assigned, it was very difficult to change track. IQ tests capture some elements of intelligence, but scores are also affected by unrelated socio-economic factors. Haynes argued that the tracking system led to discrimination and was in 'direct opposition to the American ideal'.[5]

Haynes served on the DC school board and advocated the removal of the tracking system. She spoke before the board in 1964 and criticized the system; it served 'in apartheid-like fashion to separate the underprivileged'. Haynes was elected president of the school board that same year and the tracking system was removed in 1966 after the court case *Hobson v. Hansen*. The case concluded that the tracking system deprived many black and poor students of an equal education, while favouring middle-class white students, and should be abolished.

Euphemia Lofton Haynes and her husband Harold.

How to win in a duel

Cox, Haynes and Woodard all wrote Ph.D. theses and then shifted their focus to teaching and improving conditions and access to education for students of colour. This made it possible for future Black mathematicians to have wider research careers in mathematics.

David Blackwell aspired to be an elementary school teacher and studied at the University of Illinois at Urbana-Champaign. He excelled in mathematics and completed a Ph.D. in the subject in 1941 at the age of only twenty-two. At some point, he decided to pursue research rather than teaching. Years later, he said in an interview that he had never really been interested in doing research per se: 'I'm interested in *understanding*, which is quite a different thing. And often to understand something you have to work it out yourself because no one else has done it.'[6]

However, barriers were placed in his way. Blackwell entered the Institute for Advanced Study (IAS) at Princeton on a one-year fellowship. Fellows were typically also made honorary faculty members of Princeton University, but Princeton had never even had a Black student, let alone a Black faculty member. Princeton complained to the IAS that it was abusing its hospitality by appointing a Black fellow. Much of this was kept from Blackwell by protective colleagues. 'Apparently there was quite a fuss over this, but I didn't hear a word about it,' he later said.[7]

Finishing the fellowship, he sent out over a hundred applications in search of another job. He also drove around and visited thirty-five colleges in person to ask if there was anything available. He received job offers from just three. He took a position at Southern University, and stayed for a year. UC Berkeley had initially looked promising, the head of the mathematics department, Griffith C. Evans, trying to convince the president of the university that it would be a good decision to hire Blackwell. But Evans later reneged, on the grounds that Blackwell's presence would make things difficult when he and his wife had members of the department over for dinner. His wife had allegedly said she would not 'have that darky in her house'.[8]

Then, Dudley Woodard offered him a permanent position at Howard University. Blackwell was particularly happy about this, later recalling his appointment: it was 'the ambition of every black scholar in those days to get a job at Howard University'.[9] He soon became a proficient research mathematician, and published twenty research papers in just a few years on a variety of topics. Not yet thirty, he was promoted to professor by 1947. During the summers, he sought stimulation beyond the confines of academia and became a consultant to the RAND Corporation, a political think tank. It was here that he met statistician Meyer Abraham 'Abe' Girshick and, arguably, did some of his most important mathematics.

The duo spent a lot of their time studying decision-making in duels, a subject mathematicians had previously engaged with. Nineteenth-century French mathematician Évariste Galois, who developed a field of algebra now known as Galois theory, famously participated in a duel when he was at the height of his mathematical powers. The day before, he had stayed up all night sorting out his mathematical papers and writing letters to his colleagues outlining further ideas. When morning came, he had left an impressive array that mathematicians would follow up on in the coming years. This was of little use to him on that day, however; he was shot in the abdomen and died shortly afterwards.

Blackwell and Girshick were among the first to mathematically analyse the tactics in a duel, trying to determine when the optimum moment to fire was. They took the rules of the duel to be:

- Each person has a gun loaded with a single bullet.
- Upon hearing the call of 'Fire!' the duellers can walk towards each other.
- They can fire at any time after the call.

The chance of winning depends on timing and accuracy. Try to shoot too soon and you risk inaccuracy, but wait too long and you risk being shot before you fire. Situations like these are considered zero-sum games – your loss is your opponent's gain, and vice versa.

Blackwell and Girshick's study became known as the 'Duellists' Dilemma'. The mathematics they developed can be seen as a more sophisticated version of work previously done by Gombaud and Pascal. However, instead of a simple game of chance, strategy now had to be taken into consideration. Similar lines of thinking became popular as a means to understand other high-stake scenarios between adversaries, for example between parties in the Cold War.

Twelve years after his initial attempt to become a faculty member at UC Berkeley, Blackwell eventually joined as a professor in its new department of statistics, which had just split from the mathematics department. He became a full professor in 1955 and chair of the UC Berkeley statistics department the following year. At Berkeley, Blackwell became one of the pioneers of a new area of mathematical research known as information theory, which came hot on the heels of the invention of the transistor in 1947 and the dawn of the information age.

Perhaps best known in this field is Claude Shannon, who is often dubbed the father of information theory. Among many things, Shannon came up with a measure for the amount of information in a message, defining it as the number of binary digits – 'on's or 'off's – required to encode it. In other words, the information in a message is the number of bits needed to store it. This simple idea became known as Shannon entropy, highlighting its similarity to entropy in thermodynamic physics, a measure of the different possible combinations a group of atoms can have.

Shannon found that information sent through a communications channel could be measured similarly. Given certain criteria, he managed to mathematically describe the maximum possible transmission rate for a channel, and how to use some of this rate to correct

Blackwell, teaching at UC Berkeley (*left*) and a portrait of him (*right*).

potential errors that could accumulate. Blackwell learned of Shannon's transmission capacity theorem and investigated how it played out in a whole class of different types of channels, including a particular form of communication involving multiple channels now known as the Blackwell Channel.

He didn't stop there. Blackwell soon applied his skills to other areas of information theory, as well as to a type of computer programming known as dynamic programming, along with more abstract mathematics. He was prolific, publishing over ninety papers and books throughout his life. He received twelve honorary doctorates and served the American Mathematical Society as vice president. In 2014, four years after Blackwell's death, Barack Obama, then President of the United States, posthumously awarded him the 2014 National Medal of Science Award – the country's highest distinction for scientists. It was in recognition of his incredible achievements, made possible by pioneers like Bannaker, Cox, Woodard and Haynes.

14.   Mapping the Stars

The year was 1800 and the Celestial Police were searching for a missing suspect. Hungarian and German astronomers Franz Xaver von Zach and Johann Hieronymus Schröter had formed a group of more than twenty scientists from across Europe as a sort of Neighbourhood Watch for the cosmos. Their missing suspect was a planet they believed sat somewhere between Mars and Jupiter.

This belief started when William Herschel discovered Uranus in 1781. He was one half of an astronomy duo with his sister Caroline Herschel. Together, they discovered many celestial objects, though she wasn't there for this particular one. The discovery thrust a mathematical law that had been proposed decades before into the spotlight. The Titius–Bode law stated that planets orbit the sun at locations that double in distance the further away they are. The discovery of Uranus was in accordance with this series, but if the law was to hold, there should have been a planet between Mars and Jupiter too. And the celestial police were determined to find it.

However, they had barely started their search when they were beaten to it – or, at least, it seemed that way. Italian priest and astronomer Giuseppe Piazzi spotted what he at first thought was a comet but soon came to believe was the missing planet, and named it Ceres. But Ceres was too small to be a planet; it was in fact one of those newfangled asteroids that the Herschels had been discussing.* After this initial setback, the celestial police resumed their search. In the following years, several more asteroids were found in similar locations, including Pallas, Vesta and Juno,† but no planets.

The celestial police eventually came to the conclusion that there

---

* Although it was classified in 1867 as an asteroid, it would be reclassified again in 2006 as a dwarf planet.
† All still asteroids . . . for now.

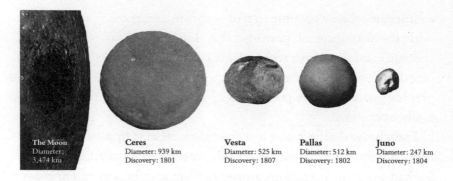

The Moon
Diameter:
3,474 km

Ceres
Diameter: 939 km
Discovery: 1801

Vesta
Diameter: 525 km
Discovery: 1807

Pallas
Diameter: 512 km
Discovery: 1802

Juno
Diameter: 247 km
Discovery: 1804

was no planet between Mars and Jupiter, but there was an asteroid belt. Many thought that the belt must be the remnants of a destroyed planet, which would have meant that the Titius–Bode law still held. Years later the discovery of Neptune in 1846 in the 'wrong' place shattered the law.

The search for a planet between Mars and Jupiter may have been in vain, but the search itself made two things clear. First, astronomers needed more data. There was plenty more to discover out in the solar system and beyond – if only they looked for it. And second, they needed to use mathematics, to analyse data gathered from observation and new photographic techniques. Recording and analysing astronomical data would become one of the most important scientific developments of the twentieth century and kick off a hundred-year international collaboration to map the skies and a trip to the moon. And at the centre of it all would be a new job – that of 'human computer'.

Look up

Throughout history, many civilizations and cultures have produced impressive maps of the night sky. The earliest copy of such a map still in existence dates from the Tang dynasty in China, around the ninth century, but Chinese astronomers were mapping the positions of the stars at least 1,500 years before that. Ancient Babylonian, Maya, Arab and Greek mathematicians and astronomers spent much of their time tracking and tracing the night sky. In the nineteenth century, it became possible to truly capture a moment in the universe with the invention of photography. Initially, it was an extremely specialist and expensive technology, but by the turn of the twentieth century and with the development of cheaper and easier photographic techniques the possibilities for photographing the night sky really opened up.

French naval officer and director of the Paris Observatory Ernest Mouchez saw the potential of these new photographic techniques. He had spent the preceding decade surveying the coasts of Paraguay, Brazil and Algeria, and observing the transit of Venus from Île Saint-Paul in the Indian Ocean. During this time, he learned about dry-plate photography, a technique that simplified many of the most complicated parts of photography for photographers. In 1882, the Great Comet became visible in the sky, initially in the Cape of Good Hope and the Gulf of Guinea, where it was so bright it could be seen during the day. The comet was also visible from the northern hemisphere, but not as clearly, and this got Mouchez thinking.

Mouchez teamed up with David Gill, a Scottish astronomer based in South Africa, and in 1887 announced a new project at the Paris Congress, an international astronomy conference, to use dry-plate photography to photograph and chart the entire night sky. This plan was more ambitious in scope than anything that had previously been attempted, and so Mouchez and Gill canvassed for help from around the world at the congress. Twenty-two observatories answered the call, including those from Europe, Africa, Asia and South America.

There were two related components to the project, the *Catalogue Astrographique* (Astrographic Catalogue) and the *Carte du Ciel* (Map of the Sky), and the aim was to photograph and map every star brighter than around magnitude 11.0. The magnitude scale is somewhat counter-intuitive, with lower and negative numbers representing objects that are brighter than higher numbers. The North Star has a magnitude of around 2. So, allowing for the quirks of the scale, that means the project was looking to catalogue every star with a brightness of $\frac{1}{3981}$ of the North Star.[*]

[*] We said it was weird!

The observatories used similar glass plates, telescopes, magnifiers and exposure times to try to ensure uniformity when measuring their particular parts of the sky. Each photograph was around 13 × 13 centimetres in size and captured a small 2.1° × 2.1° segment of the sky. The aim was for the project to obtain the data in about ten to fifteen years.

In the end, it took something closer to a hundred years. The millions of stars and celestial objects photographed had to be analysed manually and human computers were gathered to help analyse the data. The people working on these projects became skilled astronomical mathematicians, with many of them developing techniques that would later be used in astronomy.

Harvard computers

An astronomer, Edward Charles Pickering, and the Harvard College Observatory he ran would have been a big asset to the *Catalogue Astrographique* and the *Carte du Ciel*. Pickering had become famous for his remarkable images of the moon, capturing it in such detail that his images looked more like paintings made by an alien than a photograph taken from Earth. As director of the observatory, he had access to the largest telescope in North America at the time and, following a donation, he found himself in charge of a second powerful telescope too – but this had left him with a bit of a problem.

The telescope had been gifted to the observatory by Mary Anna Palmer Draper. She was a wealthy amateur astronomer, as was her husband, Henry, who had been the first person to photograph the moon in 1840. When Henry died suddenly, Mary donated their telescope to the observatory and set up a memorial fund to help pay for the work done there. Pickering and his team needed no encouragement and took a huge number of photographs using it, but soon fell behind on the analysis. In total, they had 120 tons of photographic plates of the sky that needed to be processed.

Pickering had to hire more people to help with the processing. Among his staff at home was a woman called Williamina Fleming, who was working as a maid after her husband had abandoned her, leaving her to support their young son alone. Pickering's wife, Elizabeth, noticed that Fleming was extremely sharp and suggested that she could help. Fleming learned the basic tools of astrophotography analysis from Pickering and began working at the observatory as a paid computer — though she was paid half of what a man in a similar position would have been. She soon excelled in the role and took on further responsibilities, helping to oversee the team as curator of astronomical photographs. As a result she became the first woman to hold a formal title at Harvard University.

Most photographs and recordings were still taken by male astronomers, but the mathematical analysis became women's work. Pickering would go on to hire around eighty female computers. Cost was one of the primary factors that drove his decision; nevertheless, the Harvard computers became pioneers in the field of astronomy, developing important methods to analyse and classify what they saw. There were patterns to be found across the images and these women set about finding them.

Antonia Maury, a niece of Mary Anna Palmer Draper and Henry Draper, was among those to join Pickering and Fleming's team. Maury had learned astronomy at college from the first American to discover a comet, Maria Mitchell.* She knew how to catalogue celestial objects quickly and brought that knowledge to the observatory. One of Fleming's first tasks at the observatory had been to improve the way stars were classified based on the wavelengths of light – spectra – recorded from them. These wavelengths were recorded with a spectroscope, an instrument that splits light from the telescope into its component wavelengths. The presence or absence of different wavelengths acts as a sort of chemical signature for the star, allowing astronomers to work out the elements present. Fleming came up with a system – the Fleming–Pickering system – for classifying stars based on the amount of hydrogen they contained.

Maury used spectra in her work too. Pickering discovered the

---

* The first woman in record was Maria Margaretha Kirch in Berlin in the early eighteenth century.

first binary star – a pair of stars gravitationally bound to each other – and asked Maury to work out what its orbit was. By analysing the patterns in the spectra, she was able to calculate how the binary star moved.

Annie Jump Cannon joined the group and reinvented the way stars were classified based on spectra. She had got the astronomy bug from her mother, Mary Jump. In Annie's childhood, the pair had used an old astronomy textbook to identify constellations they viewed from their attic. Mary encouraged her daughter to attend Wellesley College in Massachusetts to study physics and astronomy, which she did, before becoming one of the Harvard computers. Jump Cannon personally categorized 350,000 stars – more than anyone else – and took over the direction of the team after Pickering's death. Her most lasting legacy was to incorporate star temperature into the Fleming–Pickering

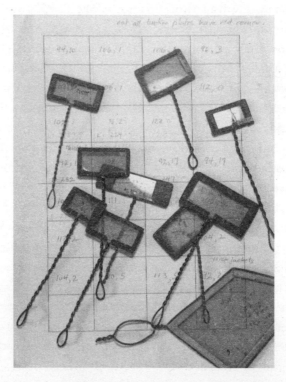

Glass plates were used by the Harvard computers as tools to compare the brightness of different objects captured on photographic plates.

classification system. This new system became known as the Harvard Classification Scheme, which is still in use today.

Another member of the team, Henrietta Swan Leavitt, discovered that particular stars called Cepheid variables pulsate in relation to their brightness, an observation known as Leavitt's law. Leavitt's law makes it possible to work out the actual brightness of a Cepheid variable, and this, when compared with how bright the star appears, can be used to determine how far away it is. Cepheid variables are now seen as one of the most important indicators of galactic and extra-galactic distances, and the formulation of Leavitt's law has led to a greater understanding of the universe, among them its size.

Gösta Mittag-Leffler, the same man who had helped Sophie Kowalevski, intended to nominate Leavitt for a Nobel Prize for her discovery; unfortunately, she died before he submitted the nomination. When he heard of the proposed nomination, the then director of the Harvard Observatory, Harlow Shapley, wrote to Mittag-Leffler, attempting to take the credit. This was an all too familiar story to the Harvard computers, who have sometimes derogatively been called 'Pickering's Harem'. They were very rarely given any credit in publications of the Harvard College Observatory. Maury left in 1891 in protest, saying when Pickering asked her to return: 'I do not think it is fair that I should pass the work into other hands until it can stand as work done by me. I worked out the theory at the cost of much thought and elaborate comparison and I think that I should have full credit for my theory.'[1]

Maury did return to the observatory, and published a catalogue of classifications in 1897 entitled 'Spectra of bright stars photographed with the 11-inch Draper telescope as part of the Henry Draper Memorial and discussed by Antonia C. Maury under the direction of Edward Charles Pickering'. This was the first publication by the Harvard College Observatory to have a woman's name included as an author. Yet despite Fleming having been given her official role as curator in 1899, it was only in 1908 that her name was included in any of the catalogues of stars and their spectra produced by Harvard.

It's hard to deny the contribution of the Harvard computers. They documented and analysed millions of stars and rewrote the rule book

Processing data at Harvard Observatory (*top*); Pickering and computers (*bottom*).
Women were usually barred from making telescopic observations,
although some exceptions were made.

for how they should be categorized. Although they never officially
joined the project of the *Catalogue Astrographique* and the *Carte du
Ciel* – Pickering believed that it was simply too ambitious and too
time-consuming for astronomers – Mouchez and Gill's project would
adopt the techniques and standards developed by the Harvard team.

> VACANCIES exist at Royal Obser-
> vatory for girl computers, J.C.
> standard: commencing salary £9 p.m.
> plus COLA at present £23/16/8;
> hours 9-1 and 2-3 30; neatness and
> accuracy in figures essential.–Apply
> in writing to Secretary.   8031

A South African advert for 'girl computers' from the 1940s.

Just a few more photographs

The Harvard computers were not the only human calculators and analysts. Similar groups of people were recruited to work on the *Catalogue Astrographique* and the *Carte du Ciel*. Some were women, such as those at Gill's observatory in South Africa, though very few details about them were recorded. At the Vatican's observatory, a team of nuns performed the analyses, and there were many women at the Adelaide, Sydney, Melbourne and Perth observatories. However, even with huge numbers of human computers, Mouchez and Gill's ambitious project spiralled out of control. The first photographs were taken in 1891, the last would not be taken until 1950, and even then many of these images remained unprocessed for years.

Much of the delay was caused by the project simply being too large in scope. It also fell prey to the consequences of world events. In Mexico, for example, architect, engineer and the first director of the National Astronomical Observatory, Ángel Anguiano, was invited to participate in the project, but then civil war broke out in

Anguiano set up the National Astronomical Observatory of Mexico.

1910, leading to the Mexican Revolution. In the same period the First World War meant that there were delays in getting the supplies needed to take the images.

The *Catalogue Astrographique* was finally completed in 1987, a hundred years after the project began. Initially, in pure scientific terms, it was far from a success – by the time it was finished, more advanced photography and analytic methods had already been developed – but it did pave the way for future ambitious global projects. The *Carte du Ciel* was never completed.

In the 2000s, there was a revival of interest in the data and images from both aspects of the project. The printed catalogues were converted into a machine-readable format, creating a rich store of historical data, and, by using comparisons with more modern star catalogues produced using satellite telescopes, this data has been used to help derive the motion of millions of stars.

The race for space

Human computers existed before this astronomical boom. For decades astronomers and mathematicians had hired assistants to perform calculations. Eighteenth-century scientist Alexis Claude Clairaut worked out how to allocate portions of the calculations so that they could be done in parallel, kickstarting a movement where many different people could work towards an answer at the same time. The paths of many celestial objects were recorded in this way. The United States Signal Corps put together a team that calculated weather patterns by working in intensive two-hour shifts. Initially, women were rarely employed as computers, but this started to change in the mid- to late nineteenth century. By the twentieth century, the vast majority of computers were women.

Teams of computers sprang up for everything from performing statistical analyses to calculating the stresses on the Afsluitdijk dam in the Netherlands. However, it was the two world wars that really ramped up the demand for computers and turned the role into more of a profession. Computers on both sides worked through lengthy calculations to produce navigation tables and map grids, analyse codes and predict the paths missiles would take.

When the space race began between the US and the USSR, computers were once again needed to perform important calculations. As early as 1951, a team of female mathematicians designed the first digital computer in the Soviet Union. Female computers in the USSR also worked on calculations for astronomical catalogues, processing spectral information and creating data banks. Pelageya Shajn, for example, discovered over 150 new variable stars, in addition to 40 new minor planets and a comet, which was named after her. Sofia Romanskaya carried out 20,700 observations about the rotation of the Earth. Evgenia Bugoslavskaya wrote a textbook on photographic astronomy, and Nadezhda Sytinskaya wrote up observations of the moon.

On the US side, and celebrated in the book and 2016 film *Hidden Figures*, a group of African American women were similarly central to the space race, with Katherine Johnson at the forefront. She started her career at the National Advisory Committee for Aeronautics, which later became part of NASA. She was initially a member of a team that performed calculations related to aeroplane black boxes but one day was temporarily assigned to assist the all-male, all-white flight research team. Her knowledge of mathematics and geometry was so strong that she was permanently assigned to the Guidance and Control Division. Here she helped work on the calculations that were used to put Alan Shepard into space and John Glenn into orbit – the first Americans to achieve each feat* – and for the paths that the Apollo missions would take to the moon.

Rocket science, in principle, shouldn't be that difficult. Writing out the equations that govern a rocket going into space or around the moon is fairly simple, using the laws of gravity and motion. However, actually solving the equations in many instances turns out to be practically impossible. Numerical methods have to be used to approximate the solutions. For a trip to the moon, you need thousands and thousands of approximations along the trajectory to be sure that

---

* Yuri Gagarin from the Soviet Union was the first person in space and in orbit. Valentina Tereshkova was the first woman in space.

nothing will go wrong. As Katherine Johnson said, 'They were going to the moon. I computed the path that would get you there. You determined where you were on Earth, when you started out and where the moon would be at a given time. We told them how fast they would be going and the moon would be there by the time they got there.'[2]

One of the numerical techniques in use was Euler's method. The principle at its heart is simple – rather than trying to solve the overall differential equation, calculate its values at some specific points and connect the dots. The result is an approximation of the actual solution, but the smaller the gaps between the points, the closer it is to the true answer. However, herein lies the trade-off. Accuracy is obviously important when travelling to the moon – you don't want to miss! – but each calculation takes time, and if too much accuracy is required, it could take decades to complete the analysis.

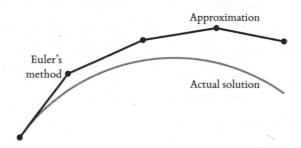

Euler's method gets closer to the actual solution with each additional calculation.

It's no wonder that when a new technology came about that promised to vastly speed up the calculations, it was quickly adopted. When IBM 7090s arrived, the company said it could perform hundreds of thousands of additions and subtractions in a single second. Used correctly, this would greatly accelerate the calculations needed to get people into space. Many of the female computers realized the importance of the new technology and reskilled as computer programmers. Evelyn Boyd Granville at IBM, for example, worked on many of

NASA's calculations, including those made to work out the entry point and launch window for John Glenn's first orbit. However, some people were suspicious of the technology. Before flying, John Glenn famously demanded, 'Get the girl to check the numbers.' The 'girl' being Katherine Johnson.

Mission control during John Glenn's first orbit.

Mathematics is very much a tool when it comes to space flight. Approximations are good enough, so long as you know the margins for error. Never do you need to hit an exact target. Maybe the wiggle room is just a few millimetres, maybe it's more. Though precision is important, perfection is not. However, this isn't true in much of mathematics, where the whole point is to discover exact relationships between specific things. These might be properties of shapes and mathematical functions, or even the point on the number line where something happens. The exact number for such a thing can be mathematically important, even if it is mind-bogglingly massive.

15. Number-crunching

$10^{10^{10^{34}}}$ is so large a number it is virtually impossible to comprehend. It begins with a one and is then followed by a lot of zeroes, but if somehow you managed to write down a zero for every atom in the universe, you would still not be close to having written out this number. Not by a long shot.

Suppose you had started the task of writing out $10^{10^{10^{34}}}$ at the dawn of time, as the universe sprang into existence. And suppose that you were extremely proficient at writing zeroes. So much so that once for every second that has passed since the Big Bang you had added the same number of zeroes to your tally as there are atoms in the universe. Unfortunately, even then you still wouldn't be close to finishing writing this number. In fact, it would be fairer to say that you wouldn't even have really got started.

'OK,' you might think to yourself, 'how far have I got?'

Well, if you wrote your progress so far as a fraction, it would be one over a number so large that it would have more digits than there are atoms in the universe. In fact, if you had started writing this number at the dawn of time . . .

$10^{10^{10^{34}}}$ is unfathomably large. It fits into a branch of mathematics known as number theory. At its core, number theory is exactly what it sounds like: the pursuit of a theory that underpins numbers. But numbers can be weird. Learning the basics is one of the easiest things you can do in mathematics, but when you dive a bit deeper numbers often become unreasonably complicated. There is probably no mathematical subject that has more of a disparity between the simplicity of some of the ideas and conjectures and the difficulty in proving them. This has often made number theory problems appear seductively straightforward but, as generations of ensnared mathematicians can attest, nothing could be further from the truth.

Take the Goldbach conjecture. This conjecture, proposed by a

German mathematician, Christian Goldbach, in 1742, is based on a pattern so easy that almost anyone with a bit of mathematical curiosity could have spotted it: every even number greater than two appears to be the sum of two prime numbers. For example, 8 is equal to $3+5$; 90 is equal to $7+83$ and $123,45,678$ is equal to $31+12,345,647$. Every even number anyone has ever checked fits this pattern. The conjecture – that every even number greater than two really is this way – has been knocking around for nearly 300 years, yet nobody has been able to prove it. This leaves open the possibility that there is an extraordinarily large even number that can't be split into two primes. And that would just be typical number theory – something that appears true until you find an enormously large counter-example.

The ten-to-the-ten-to-the-ten-to-the-thirty-four number above turns up this way, though it isn't about the Goldbach conjecture but about a more intrinsic property of prime numbers that we will come to later in this chapter. When number theorist Stanley Skewes found it in the 1930s, mathematician G. H. Hardy described it as 'the largest number which has ever served any definite purpose in mathematics'.[1] The number even made it into the *Guinness Book of Records*.

A large number like this that serves some mathematical purpose was destined to come out of a subject like number theory. And it fits into one of the most exciting and expansive periods in number theory history, a period that is anchored in Oxford and Cambridge in the UK but also drew in mathematicians from elsewhere in some of the greatest mathematical collaborations in history.

The mathematician known as Hardy–Littlewood

Godfrey Harold Hardy, known as G. H. Hardy, and John Edensor Littlewood were formidable mathematicians in their own right. But it was together that they had most of their success, publishing more than a hundred papers across number theory and mathematical analysis (which includes the study of the limits, differentials and integrals we discussed in Chapter 8). Before Hardy and Littlewood, mathematics in England was stagnant, especially compared to mainland Europe, but together they managed to turn things around.

The pair began their collaboration in 1910, and their characters could not have been more different: Hardy was outgoing and gregarious, Littlewood so reserved he didn't even attend academic conferences. Littlewood was so inconspicuous, in fact, there were some people who doubted he existed at all. Littlewood experienced depression throughout his life, which contributed to his withdrawn nature. Together, Hardy and Littlewood's output was both varied and significant.

Though the mathematical duo couldn't actually prove what came to be known as the first Hardy–Littlewood conjecture,★ conjecturing is an incredibly important part of mathematics. A decent conjecture can focus minds, give rise to new tools or ideas and turn the task of solving one problem into another. You get interesting answers only when you formulate interesting questions. The first Hardy–Littlewood conjecture is about prime numbers: a number that is only divisible by 1 and itself. So 5 is a prime number because it can only be divided by 5 and 1, but 6 is not, because it can be divided by 1, 2, 3 and 6. Prime numbers are incredibly important in mathematics, and especially in number theory.

The first Hardy–Littlewood conjecture gives an estimate for how often pairs of prime numbers occur that are a fixed number, k, apart. So, for example, in the case $k = 2$, the conjecture estimates the distribution of prime numbers that are two apart, such as 3 and 5; 5 and 7; and 11 and 13. These are known as twin primes, and we think there are probably infinitely many of them, but we don't actually know for sure. Solving the first Hardy–Littlewood conjecture would solve this twin prime conjecture, which has stood undefeated for over 150 years. (As a quick aside, prime numbers that are four apart (such as 3 and 7) are called cousin primes, and prime numbers that are six apart (such as 5 and 11) are called sexy primes. Sexy = sixy. On behalf of all mathematicians: apologies.)

Another of their important contributions is what is now known as

---

★ There is also a second Hardy–Littlewood conjecture. The two conjectures were announced at the same time, but it is thought that if one of them is true, then the other is false.

the Hardy–Littlewood circle method. It's a strategy for working out the behaviour of certain mathematical functions. In Hardy and Littlewood's case that strategy was the partition function.

Partitions simply investigate how many ways a number can be written as a sum of integers. The number 4, for example, can be written as $4; 3 + 1; 2 + 2; 2 + 1 + 1;$ or $1 + 1 + 1 + 1$. So we say that 4 can be partitioned in five distinct ways. The main idea of the Hardy–Littlewood circle method is to translate questions involving additions into ones involving exponentials. The approach proved not only to be novel and effective but extremely powerful too. After its announcement, number theorists flocked to learn the technique and, wherever they looked, they found uses for it. Hundreds of papers have since been written that utilize the method, including the proof of a weaker but related version of the Goldbach conjecture.

Such was their success as a duo that Harald Bohr, mathematician, international football player and brother to Niels, summed it up in 1947 by saying that there were three great mathematicians in England: 'Hardy, Littlewood, and Hardy–Littlewood'.[2]

The rules of the Hardy–Littlewood collaboration were agreed in the form of axioms. The first axiom was that when one wrote to the other, it didn't matter whether what they wrote was right or wrong. The second was that there was no obligation to read or reply to each other's letters; the third that they should try to avoid working on the same detail; and the fourth that all papers would appear under both their names, no matter what. It's good to have rules so clearly marked out and it's hard to dispute the results, but we, the authors, can't help but feel that had we followed the first and second axioms in writing this book, it is highly improbable that you would be reading it right now.

Hardy and Littlewood's ability to collaborate also extended to others, especially when they welcomed a newcomer to Cambridge from India.

Mathematics straight from God

In 1913, Hardy received an unusual letter. Postmarked India, it was written by a twenty-three-year-old clerk in the accounts department of the Port Trust Office in Madras and described a peculiar predicament he found himself in. The writer explained that he had little in the way of formal mathematics education but that he was pursuing the subject nonetheless. 'The results I get are termed by the local mathematicians as "startling".' However, local mathematicians were unable to understand his 'higher flights', he wrote, and so he was unsure how to proceed. After a few paragraphs of mathematical explanation, the letter was signed 'Yours truly, S. Ramanujan'.[3] Attached were almost a dozen pages expanding on his ideas – and they were extraordinary. Hardy said that it was 'certainly the most remarkable [letter] I have ever received . . . one thing was obvious, that the writer was a mathematician of the highest quality, a man of altogether exceptional originality and power.'[4]

Srinivasa Ramanujan's early life was marked by tragedy. He was born in Erode in the south of India in 1887 and at the age of two contracted smallpox which left scars all over his body. By the age of six he had lost two brothers and one sister, and a year later he lost his grandfather to leprosy, and the extreme stress caused Ramanujan to develop itching and boils – conditions that would recur throughout his life. Mathematically speaking, however, it was a different story. By the time he was eleven, Ramanujan was excelling in mathematics. Two local older college students who were lodging at his home found this out when they realized there was nothing they knew that they could teach him. By thirteen, he was discovering theorems of his own in trigonometry. This was even more remarkable given the current state of education: at that time, even among school-educated men of Ramanujan's Brahmin caste, only about 11 per cent were literate.

Later in his teens he received a university scholarship, but as he devoted all his time to studying mathematics, he failed in his other subjects (among them English, Greek, Roman history and physiology) and his scholarship was cancelled, which meant he did not receive a university education. Undeterred, he continued to pursue mathematics on his own. He began to pose and solve problems in mathematics journals. One of his papers on Bernoulli numbers was accepted for publication by the *Journal of the Indian Mathematical Society* in 1911, and he started to gain recognition in India for his work. Eventually, he got a job in the accounts department and started his correspondence with Hardy. Ramanujan put forward in his letter a mixture of ideas, often containing a combination of the old, the familiar and the downright surprising.

Take this equation, which appeared on one of the pages:

$$\int_{0}^{\infty} \frac{1+\left(\frac{x}{b+1}\right)^2}{1+\left(\frac{x}{a}\right)^2} \cdot \frac{1+\left(\frac{x}{b+2}\right)^2}{1+\left(\frac{x}{a+1}\right)^2} \ldots dx = \frac{1}{2}\pi^{\frac{1}{2}} \frac{\Gamma\left(a+\frac{1}{2}\right)\Gamma(b+1)\Gamma\left(b-a+\frac{1}{2}\right)}{\Gamma(a)\Gamma\left(b+\frac{1}{2}\right)\Gamma(b-a+1)}$$

Ramanujan was clearly doing calculus – Hardy could tell that from the use of Leibniz's elongated S notation on the left-hand side. But the formula and the symbols on the right were completely unexpected, producing a result Hardy had never seen before. Hardy told Littlewood about the letter and together they examined the mathematics, becoming increasingly impressed by Ramanujan's ability and originality as they worked their way through it. In his reply, Hardy asked Ramanujan to send more, believing that because the work was so spectacular he must have known the theorems underpinning it that he hadn't put on the page. Hardy described him as 'keeping a great deal up his sleeve'.[5]

Hardy invited Ramanujan to Cambridge so they could work together. Although he was initially against the proposal, Ramanujan finally relented and joined Hardy and Littlewood in England, leaving his young wife behind with his parents and bringing notebooks filled with ideas. He had already sent Hardy over a hundred theorems in his letters, but there were many more that he had yet to share. Across his work, a few of the results were wrong, some were already known, but the rest were new discoveries. Over the course of five years, he published his findings in collaboration with Hardy and Littlewood.

Just as Hardy and Littlewood had contrasting personalities, so did Ramanujan and Hardy–Littlewood. Ramanujan relied heavily on intuition. He was religious and believed that many of his insights came directly from his family goddess, Namagiri. This led to him sometimes skipping steps in proofs and derivations, in the belief that he already knew the results were correct. Hardy and Littlewood would fill in some of the gaps – and the method worked. Ramanujan regularly managed to discover completely surprising formulae, with perhaps one of the most famous being the following series for pi.

$$\frac{1}{\pi} = \frac{2\sqrt{2}}{9{,}801} \sum_{k=0}^{\infty} \frac{(4k)!(1{,}103 + 26{,}390k)}{(k!)^4 \, 396^{4k}}$$

The numbers on the right-hand side seem to come out of nowhere and to have nothing to do with pi itself. Yet this is one of the fastest series known for approximating pi. The large $\sum$ means that this is a summation, where you calculate the value for the expression when $k = 0$ then add this to the value for the expression when $k = 1$ and $k = 2$, and so on. After computing just a few terms in this series, you quickly get an extremely accurate value for pi. So much so that it still underpins several algorithms for calculating the number to many decimal places in use today.

Srinivasa Ramanujan.

Ramanujan's life was sadly cut short. He had experienced ill health throughout his life and in his early thirties contracted tuberculosis and was diagnosed with a severe vitamin deficiency – he ate a restricted vegetarian diet that was not well catered for in Britain. He returned to India a national hero, but died aged just thirty-two in 1920.

Even today, we are only just beginning to understand the impact of Ramanujan's mathematics. Probing his work in later decades, Jean-Pierre Serre and Pierre Deligne pioneered an important tool in number theory called Galois representations, which won them Fields Medals, one of the most prestigious awards in mathematics.* Advances sparked by this ultimately led to the solving of Fermat's last theorem, a conjecture that had stood for over 350 years before it was proved by Andrew Wiles in 1995. Just a few months before he died, Ramanujan wrote one final letter to Hardy in which he outlined a new theory of unusually symmetric equations known as mock theta functions. These remained mysterious for decades but are now starting to be used in string theory and the description of black holes.

This pattern is a well-established one in mathematics. Probing one thing in one area leads to results that are useful elsewhere. And this was particularly true for one of Hardy and Littlewood's lesser-known collaborators, Mary Lucy Cartwright.

From collaboration comes chaos

At the beginning of 1938 the UK government realized that it needed some mathematical assistance. War was on its way in Europe and the development of what was soon to become radar had hit a snag. The engineers at the Department of Scientific and Industrial Research (DSIR) suspected there might be problems with the equations for high-frequency radio waves, so they wrote to the London Mathematical Society looking for assistance. Of course, radar was top secret, so it was never mentioned in the memo, but, mathematically, the problem was intriguing. Cartwright was already familiar with similarly 'objectionable-looking differential equations'[6] so she took up the call.

Cartwright was a rising star of mathematics. She had begun her

---

* The first ever Fields Medals were awarded in 1936.

career in 1919 by studying mathematics at St Hugh's College, Oxford. After a couple of years of disappointing exam results in the subject, she started to turn things around in her third year after attending evening classes held by Hardy. In 1923, she obtained a first-class degree from the university – the first woman to do so. She eventually became a student of Hardy's and obtained a doctorate. She then continued as a research mathematician at Girton College, Cambridge, before Hardy and Littlewood recommended her for an assistant lectureship. She went on to become a lecturer there in 1935. It was here that she spotted the government's appeal for help. She showed Littlewood the memo and they teamed up to see if they could find a solution.

Mary Cartwright (*left*), and Cartwright at the International
Congress of Mathematicians in 1932 in Zurich (*far right*).
Many of the women pictured were wives of mathematicians.

Cartwright found Littlewood an unconventional collaborator. He was most willing to speak on walks, drawing imaginary figures on walls on the way, but this was hardly the best way to share mathematical ideas. She came across the axioms for collaboration Hardy and Littlewood had laid down and they started working together via letters instead, to great effect. 'Miss Cartwright is the only woman in my life to whom I have written twice in one day,' Littlewood later wrote.[7]

The problem the UK government was having was specifically to do with the amplifiers used in radar, which would often fail. The soldiers using them thought it was something to do with the manufacturing, but Cartwright and Littlewood found that the mathematics underpinning them was wrong. Physicist Freeman Dyson later described the situation: 'The equation itself was to blame.'[8] What Cartwright and Littlewood found was that as the frequency of the radio waves increased, the equations broke down, and this meant that small changes led to erratic and unstable results. The pair didn't quite hit on a solution to the problem, but their finding helped the engineers come up with ways to compensate for these errors, rather than searching for a problem within the manufacturing process.

Much like some of Ramanujan's results, it took a while for the implications of Cartwright and Littlewood's work to be understood. Unbeknown to them, the pair had hit upon a fundamental idea that would change the face of mathematics: the idea of chaos.

This strange mathematical phenomenon is perhaps best known for the butterfly effect, which suggests that a butterfly flapping its wings in one location may be enough to set off a tornado on the other side of the planet. At its core is the study of systems that are sensitive to initial conditions. In such situations, minuscule differences in the way something starts can go on to have huge effects and make its subsequent course unpredictable in the long run. Weather is a chaotic system. Even though we can make reasonable enough suggestions about what will happen in the short term, in the longer term, making specific predictions is impossible. So even a tiny difference, such as the flutter of a butterfly's wings, can have a big impact. We now know that there are many chaotic systems, including how fluids flow, planetary systems and even the stock market.

But chaos is not random, it's just very hard to predict – and this kind of unpredictability is exactly what Cartwright and Littlewood had found, some twenty years before the development of chaos in the 1960s.

Cartwright had a long and illustrious career. She achieved many firsts for women, including being the first to serve on the Council of the Royal Society and to be president of the London Mathematical

Society. She lived to be nearly one hundred, so was able to see the incredible field she helped lay the groundwork for flourish.

Fundamental theorems and problems

Littlewood saw a principle of his in action with his work with Cartwright: attempt to solve a difficult problem; you may not solve it, but you might prove something else. He had first discovered this when he tried to take on the Riemann hypothesis. Despite all the success of Hardy–Littlewood and their collaborators, the Riemann hypothesis was one they were never able to crack. This incredibly important problem is a classic of number theory. From the simple question 'How often do prime numbers occur?' comes an unavoidable fall into a pit of complexity, confusion and, for some, despair.

Let's start with the simple(r) parts. One of the wonderful things about primes is that they are the building blocks of other numbers – a phenomenon formalized by the fundamental theorem of arithmetic. This theorem states that every number is a unique product of prime numbers. Take 15, for example, which is the product: $3 \times 5 = 15$. Both 3 and 5 are prime numbers and there is no other combination of prime numbers that will get you 15. Or try 42, which is uniquely made up of the prime numbers $2 \times 3 \times 7 = 42$. Whatever number you pick, this fundamental theorem of arithmetic tells you that when you look more closely you'll find primes. Or, as Hardy put it, 'The primes are the raw material out of which we have to build arithmetic.'[9]

So how common is this raw material? If you do a bit of exploring, it's fairly easy to work out that prime numbers become rarer as you work your way up the number line. There are twenty-five prime numbers between 0 and 100, but only twenty-one between 100 and 200 and only sixteen between 200 and 300. Logically, you might think this seems reasonable – for larger numbers there are more numbers to divide by. A reasonable thing to then ask would be: is there a formula that describes this relationship?

There is. It is called the prime number theorem and it was proved independently by French mathematician Jacques Hadamard and

Belgian mathematician Charles Jean de la Vallée Poussin. There are several different variations of the result, but one gives the following formula to tell you approximately how the primes are distributed.

$$\pi(x) \sim \int_2^x \frac{dx}{\log x}$$

On the left-hand side is an expression that tells us the number of primes below x and on the right-hand side is an expression for calculating it. The squiggle in the middle means 'is approximately equal to'. And as the table below shows, the approximation is pretty accurate.

n	Number of Primes less than n	$\int_2^x \frac{dx}{\log x}$
1000	168	178
10000	1229	1246
50000	5133	5167
100000	9592	9630
500000	41538	41606
1000000	78498	78628
2000000	148933	149055
5000000	348513	348638
10000000	664579	664918
20000000	1270607	1270905
90000000	5216954	5217810
100000000	5761455	5762209
1000000000	50847534	50849235
10000000000	455052511	455055614

Proving the prime number theorem was an arduous process. Although it was conjectured in the 1700s, in standard number theory style, it took another hundred years or so to prove. The proof relied on mathematical curiosities known as Riemann zeta functions, which were first discovered by Bernhard Riemann, who we previously met as someone exploring hyperbolic and elliptic geometries. Primes do not appear with any apparent pattern but, just as the prime number theorem gives us an insight into their usual behaviour, Riemann zeta functions give us an insight into the way primes fluctuate around the average. Riemann hypothesized that certain solutions to Riemann zeta functions lie on a nice straight line. This may seem trivial, but knowing how primes fluctuate is a big deal. If the Riemann hypothesis is true, it would mean not only that several open problems in number theory were immediately solved but would also bolster our understanding of prime numbers, showing that they fluctuate in a mathematically pleasing way.

Although his hypothesis was true for every case he tested, he couldn't find a proof. Littlewood had also attempted to find one when he was younger, but without much success – as had many other mathematicians. A strange habit formed: many mathematicians simply assumed it to be true and took it from there. Some were sceptical of the approach, but this is exactly what Stanley Skewes had to do to find his number.

Skewes's number

Have another look at the table above. If you look closely, you'll see that the approximations in the column furthest right are all overestimates. In fact, if you continued this table, you would still find that the formula is an overestimate. But as with all things number theory, the simple can become complex, and what appears to be the case may just be a deception.

In 1914, Littlewood found that the formula actually flips from being an overestimate to an underestimate infinitely often. However, nobody knew of a number where this flip occurred. That is, until Skewes arrived.

Stanley Skewes was part of the first wave of students to gain a fully South African university education when he graduated from the University of Cape Town with a degree in civil engineering in 1920 and a master's in 1922. Universities had existed in the country before this, but they had really only been outposts of British ones. This started to change in the early 1900s when a movement towards establishing a university for the purpose of teaching, as opposed to giving

Stanley Skewes.

examinations and conferring degrees, began, and led to the University of Cape Town becoming a university in its own right in 1918. Thanks to this movement, knowledge production in South Africa started to become more home-grown. South African mathematicians who travelled to England to study returned with degrees and began teaching at the higher institutions in Cape Town and Johannesburg. These universities and colleges became home to talented mathematicians, and scientific knowledge spread to the wider population through various programmes on the radio.

In the eighteenth century there was little segregation at schools in South Africa. However, this had changed by the twentieth. Discrimination on a major scale against Indigenous communities was turbocharged when the National Party gained power in 1948 and apartheid became national policy, enforced by the all-white government. The University of Cape Town was one outpost that tried to resist the policy, and it became famous for its opposition. The university had a small number of Black students enrolled in the 1920s and continued to admit them throughout apartheid, but the numbers remained low while the system was in place.

Back in 1918, the University of Cape Town was still finding its feet. Skewes discovered this first hand when he won a scholarship in 1923 to study mathematics at King's College, Cambridge. The level of knowledge required was so different he essentially had to start his studies all over again from undergraduate level. Although he did manage to finish the course at King's, he only received a second-class degree, which he found deeply discouraging. He returned to South Africa and took up a post as a lecturer at the University of Cape Town, returning to Cambridge from time to time to continue his studies, first obtaining a master's, then a Ph.D., under the supervision of Littlewood, who also grew up in Cape Town. Skewes became an associate professor at the University of Cape Town and inspired many students, among them Dona Strauss.

Strauss had entered the University of Cape Town when she was just fifteen years old. Three years later, she had a BSc and, a year after that, an MSc. There were two hundred maths students in her year, but only four were women. She later said that the other three ended their mathematical careers to become homemakers, but she had other plans and so applied to Cambridge. And who should be Mistress of Girton college when she applied there? Chaos theory's very own Mary Cartwright. Girton and Newnham were the only two Cambridge colleges that women were allowed to attend at that time, and female mathematicians were few and far between. Strauss went straight into the Ph.D. programme. She felt that she had been well prepared by what she had learned from Skewes, and she was right. She thrived in Cambridge and went on to become a professional mathematician.

In her Ph.D. thesis Strauss pioneered pointless topology, which, despite its name, is a useful branch of mathematics. Topology in general is about studying shapes, but from a different perspective to geometry. For example, instead of worrying about hard differences between shapes, topology focuses on the ways in which objects can be squished into other ones. From its esoteric beginnings, topology has found uses in everything from biology to theoretical physics. Normally, topological shapes are constructed using points, just as they would be in other areas of mathematics. But topology doesn't really care about particular points, so it's a strange set-up. What Strauss investigated was how to do away with them altogether, thus making topology point-free. She also demonstrated some of the uses this version of topology had in applied mathematics and computer science.

Strauss has produced more than two hundred papers, collaborated with many pure and applied mathematicians and co-authored three books. Her work was widely recognized when the University of Cambridge hosted a conference in honour of her seventy-fifth birthday in 2009, for which mathematicians from around the world gathered to celebrate her productive life as a mathematician.

Skewes was an inspiring teacher, and one way he captivated his students was with his work on prime numbers and the prime number theorem. It was by analysing this theorem in 1933 that he hit upon his number – the extremely large one at the start of this chapter. This number was the smoking gun everyone had been looking for – the moment when the formula for the distribution of primes became an underestimate rather than an overestimate. His result gained lots of attention and impressed Hardy. But there was a potential problem: to reach his result, Skewes had to assume that the Riemann hypothesis was true. He did later, after many years, also prove that his number held if the Riemann hypothesis was false, but some mathematicians had their doubts that this was the right approach. In mathematics, true or false are not the only options. There is a third category called 'undecidable', where the axioms of mathematics being used do not allow the result to be proven.

Another of Littlewood's students, Albert Ingham, wrote to a promising young mathematician expressing his disdain at the approach: 'Skewes is not the best person to squeeze the last drop of juice out of the lemon. I am interested in *your* efforts.' That promising mathematician was a young man named Alan Turing.

An enigma even Turing couldn't crack

It is hard to mention Turing without talking about his work during the Second World War. The day after the UK declared war on Germany in September 1939, he reported for duty at Bletchley Park, a country house that became the centre of Allied code-breaking efforts.

As the war escalated, the more important it became to intercept and decode information. On one side, Nazi Germany was using one of the most sophisticated encryption machines yet created, the Enigma; on the other was Alan Turing as head of Hut 8 and his colleagues trying to break it.

Breaking codes had been a preserve of mathematicians for years. Ninth-century Arab mathematician and polymath Abū Yūsuf Ya'qub ibn Ishaq al-Kindī recorded one of the most important techniques in around 850. He noticed that by performing a statistical analysis on encrypted text you could work out what some of the substitutions were. For example, in English, 'e' is the most common letter. Using al-Kindī's approach, if you had a block of encrypted text (where the letters in the original text have been replaced with other ones), you could expect that the most common encrypted letter should be decrypted as 'e'. Continuing in this way, you could hope to decrypt enough parts of the text to be able to guess words and then decrypt the whole text.

However, in the Second World War this approach was only so useful. The German military changed the Enigma machine settings regularly, which meant that to have any hope of cracking it, Hut 8 needed to be able to work quickly – more quickly than was humanly possible. So they built a machine to perform the calculation required. Known as the Bombe, it was huge, two metres tall and two wide, and weighed roughly a ton. It took around five months to build the first Bombe, and over the following years about two hundred were built.

Turing continued to build computing machines after the war, but he also returned to number theory and started corresponding with Skewes, who was thirteen years his senior, the pair having become good friends while Skewes was visiting Cambridge in the early 1930s. Turing would go on to become a household name in mathematics, but when he met Skewes he was still an undergraduate. On that visit, the pair found themselves sitting one in front of the other during rowing practice. They started chatting about mathematics on the boat and, in particular, about the Riemann hypothesis.

Turing remained fascinated by the Riemann hypothesis, and his first published paper after the war was on this topic (although he had written the content some years before). He wondered if he could improve on the derivation of Skewes's number using his computation skills. He was sceptical that the Riemann hypothesis was true, so he tried to build a machine that could obtain solutions of the Riemann zeta function in the hope that he would find a solution that didn't lie on the nice straight line predicted by the Riemann hypothesis and thus prove it wrong. His design consisted of eighty gears in precise ratios and a counterweight, but he had to halt the project when war broke out. After the war, he returned to this problem, but digital computers were so advanced by then that his design became redundant. He instead found a way to sidestep the Riemann hypothesis completely. In doing so, he managed both to get rid of the shaky ground that Skewes's result relied upon and to find a much better Skewes number. Skewes's original was $10^{10^{10^{34}}}$. Turing, with the aid of extensive computations, suggested that 10^{10^5} would work.

Much like a collaboration between Hardy and Littlewood, Turing wrote both his name and Skewes's name at the top of the work, 'On a Theorem of Littlewood'. However, it was never published. At this point, he was living in Manchester and was in a relationship with a man called Arnold Murray. Homosexuality was illegal at the time, but he believed that the odds of being prosecuted were one in ten. Unfortunately, in early 1952, he was arrested and charged with 'gross indecency' and forced to undergo chemical castration, given injections of nonsteroidal oestrogen to 'feminize' his body. As Jeanette Winterson wrote in her essay collection about artificial intelligence, *12 Bytes*, 'What Turing wanted to do with his body – have sex with men – seemed to be more important to petty, post-war Britain than what he could do with his mind.'

The treatment greatly affected him both mentally and physically, but he kept his academic job and continued corresponding with scholars, scientific journal editors and students. On 9 April 1953 he sent a letter to his old friend Skewes in which he apologized for encroaching on Skewes's turf.

> I feel rather guilty about having invaded the territory of your number at all. One might have supposed that it could remain a pleasant corner

one could keep to oneself. However, you made the mistake of talking to me about it from time to time when you were rowing two and I at bow until eventually I thought I had better find out what it was all about, and having done so, I could not refrain from playing at it myself.

The following year, Turing died. An investigation determined that the cause of death was suicide, although some believe it could have been accidental poisoning.

Several years later, Skewes received a letter from Littlewood, who had heard about the unpublished manuscript with his and Turing's names on it. Skewes wrote back that he had visited Turing frequently and continued their written exchanges. Skewes suggested that the joint manuscript should be published posthumously under Turing's name alone, although it never was. It is tantalizing to think what could have happened had Turing and Skewes collaborated further, especially on the topic of the Riemann hypothesis, which remains unsolved. At least, for now.

Epilogue

Mathematics and its history are constantly evolving subjects. What we have written in the previous pages reflects our current best understanding of the origins of mathematics. There is a growing appreciation that the traditional story told about it, often with too heavy a focus on a small cast of ancient Greek characters, is just one component of a far richer and far more international history.

Mathematical ideas have sprung up across the world in many different guises. There's no doubt that historians will continue to uncover more complexity in each of these stories in the coming years. Some of this will be in the form of re-examining old stories and trying to shake off the prejudices of the time or those imposed by later generations. Errors caused by making sexist assumptions or by prioritizing some groups over others will be exposed, and the truth – or as near as we can hope to get to it – will be uncovered. But we can also hope for the discovery of old texts, tablets and artefacts, as well as the application of techniques such as carbon dating and DNA analysis, to give us more clues as to what really happened.

As authors, we are aware that we, too, come with our own pre-existing ideas and biases. We have done our best to let the facts speak for themselves, but this also comes with its own challenges. What to include and not to include in a history of mathematics is to some extent an ethical decision. Who gets the credit? Who is erased by omission? No history book can ever be complete. Instead, we hope that we have bent the narrative arc of mathematics towards a fairer and more representative history, not for the sake of any ideology, but because it is the truest reflection of how mathematics has developed over millennia – and how it continues to develop. Mathematics is still full of life and is as much an international team sport as ever.

Take arguably the biggest mathematical breakthrough of the last thirty years: Andrew Wiles cracking Fermat's last theorem in 1995.

Recall that Fermat's last theorem says there are no whole numbers that satisfy the equation $x^n + y^n = z^n$, where n is greater than 2. Pierre de Fermat wrote in the margins of a book that he had a proof of this nearly four hundred years ago, but that proof has never been found. Mathematicians such as Sophie Germain, among many others, attempted to tackle it, but it took the full power of modern mathematics and a whole suite of characters to get the job done.

Wiles used an impressive array of mathematics in his 129-page proof of the conjecture but central to his approach was work by two other twentieth-century mathematicians from Japan, Goro Shimura and Yutaka Taniyama. Shimura built on work by Taniyama to notice a connection between two seemingly unrelated mathematical concepts known as elliptic curves and modular forms. He conjectured that they were one and the same thing, just viewed from two different perspectives. French mathematician André Weil popularized the Taniyama–Shimura conjecture in the West before mathematicians, among them Gerhard Frey from Germany and Ken Ribet from the US, confirmed that proving it would imply that Fermat's last theorem was true as well.

By the mid-1990s, it could be said that Fermat's last theorem was the most famous unsolved problem in mathematics. And, by building on all the work that had gone before, Wiles proved the Taniyama–Shimura conjecture and therefore Fermat's last theorem too. He had worked in secret for seven years up to this point and his initial proof contained some problems. Fellow Brit Richard Taylor helped iron them out, which led to him being named a co-author on one of the papers outlining the result.

Mathematics is a relay

Wiles ran the final leg of the race, but he could not have done it without the many mathematicians who each carried the baton towards the finish line before him. In this case, much of the collaboration happened indirectly. One person inched forward on the problem, only for another to come along and use what the previous person had done to inch it on a little further. But it doesn't always happen this way. Mathematics can be a social pursuit, too.

Paul Erdős was originally from Hungary but left during the rise of Nazi Germany and spent his life touring the world, collaborating with different mathematicians. Famously, he would turn up at a colleague's house unannounced and declare, 'My brain is open.' He would stay long enough to collaborate on a few papers before moving on to the next colleague, often asking his current host who to visit next. Over his career, he collaborated with over 500 mathematicians and published around 1,500 papers – more than any other mathematician to date.

Erdős was so prolific that some mathematicians still track how 'close' they were to having collaborated with him. Having an Erdős number of 1 means that you published work with the man himself; an Erdős number of 2 that you published work with someone who published work with Erdős; and so on.* Only one person has an Erdős number of 0, and that's Paul Erdős himself. Over eleven thousand people have an Erdős number of 2, showing just how far-reaching collaboration can be.

It's hard to gauge exactly how many mathematicians there are in the world at any one time, but the Mathematics Genealogy Project gives some hints. The project started in the mid-1990s and on its website its stated aim is 'to compile information about ALL the mathematicians of the world'. Any mission statement with such a brazen use of capitals should certainly be taken SERIOUSLY.

The numbers of entries in their records by year show that mathematicians have grown substantially as a group in recent years.

The exact numbers should be taken with a large pinch of salt. It is much easier to collect records about people completing Ph.D.s in the last few decades than a hundred years ago and it takes a while for more recent years to be fully up to date. However, for decades now, mathematical Ph.D.s recorded by the project have been in the thousands each year rather than in the tens and hundreds.

If we look beyond the graph, however, it is clear that mathematics

* Timothy has an Erdős number of 4 and Kate's is 5.

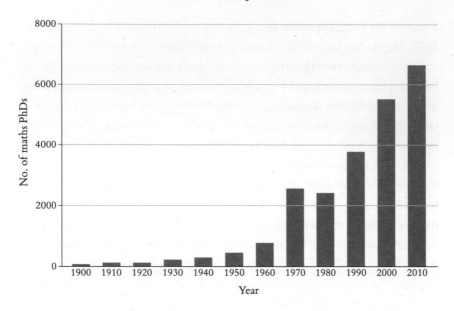

is still skewed. Although women make up around 50 per cent of Ph.D. students across all subjects[1] in OECD countries, in mathematics the proportion is far lower. According to the US Department of Education, women earned just 29 per cent of doctoral degrees in mathematics in 2013–14.[2] The She Figures 2021 report from the Publications Office of the European Union found that just over 32 per cent of doctoral graduates in mathematics and statistics in Europe were women,[3] compared to nearly 50 per cent across all subjects. Across Africa, women constitute around 30 per cent of those doing research in science, technology and mathematics.[4] These are just a few data points, but other reports tell a similar story.

Although it is happening slowly, there are small signs that this is starting to change. Take two of the biggest prizes in mathematics – the Fields Medal and the Abel Prize. Prizes are only one aspect of what is important in science, and they are often political. The history of physics is not the same as the history of Nobel Prizes in physics, for example. However, they are a reflection of what – and perhaps more importantly who – is championed.

Despite over one hundred Fields Medals and Abel Prizes having been awarded, until around a decade ago every single one had gone to a man. This changed when Maryam Mirzakhani won a Fields Medal in 2014. Fields Medals are handed out once every four years at the International Congress of the International Mathematical Union to between two and four mathematicians under the age of forty. It is arguably the most coveted prize in mathematics and winning it is an incredible recognition of someone's work. Since it was first awarded in 1936, ninety people have won a Fields Medal, eighty-eight of them men.

Mirzakhani bucked the trend in many ways. She was born in Iran and showed immense mathematical talent from a young age. As a teenager, she became the first Iranian woman to win a gold medal at the International Mathematical Olympiad, a well-established competition in which young mathematicians test their skills on the world stage. She followed this up a year later with another gold medal and a perfect score. She moved to the US to pursue a research career in mathematics, holding positions first at Princeton University and then at Stanford. She soon homed in on a field of study, focusing on strange geometric objects and spaces, building on the sort of work done by Bolyai, Gauss, Lobachevsky and Riemann. Here, along with her co-author Alex Eskin, she proved a result so magnificent it became known as 'the magic wand theorem'.

Here's how Eskin explained it in 2019.[5] Imagine a weirdly shaped room made out of mirrors with a candle in the middle. The light will bounce off the mirrors, but will it illuminate every single spot or will some points in space remain unlit? The magic wand theorem answers this. 'There are no dark spots,' said Eskin. 'Every point in the room is illuminated.' The dark spots in a candlelit room may seem like a very specific situation, and it is, but the magic wand theorem is far more general. Using concepts from algebra and geometry, it codifies a broad relationship between certain shapes and certain paths. The result is that the magic wand can be waved in many situations where there are moving particles. There are plenty of these situations, especially in theoretical physics, and so the full applicability of the magic wand theorem is still being uncovered.

Mirzakhani died young, but her story continues to inspire girls, especially in Iran, to study mathematics. Sharif University of Technology has named the library in its college of mathematics after her,

and her old school, Farzanegan high school, has named its amphi-
theatre and library after her. International Women in Mathematics
Day is now celebrated on 12 May, Mirzakhani's birthday.

In 2022, as we were making some of the final tweaks to this book,
the latest four Fields Medallists were announced. Among them was
Ukrainian mathematician Maryna Viazovska, for her proof that 'E8
lattice provides the densest packing of identical spheres in 8 dimensions'. In other words, she worked out the way to pack spheres
together in eight dimensions to leave the least amount of empty
space. How to pack spheres has been a long-standing and surprisingly
difficult mathematical problem. In two dimensions, where the
spheres are circles, Joseph Lagrange proved in the 1770s that a hexagonal pattern is best, where a single circle is surrounded by six other
circles as in a honeycomb.

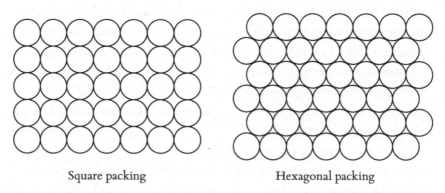

Square packing   Hexagonal packing

You can fit more circles into a tight space using hexagonal packing.

It was not until the late 1990s, more than two hundred years later,
that the next result would be proved, when Thomas Hales showed
that 'close-packed structures' – a sort of configuration that often
occurs naturally in crystals – are the optimum way to pack spheres in
three dimensions. However, the result was not yet known for any
higher dimensions – until Viazovska proved it for eight dimensions
in 2016. Shortly afterwards, she became involved in working out the
best configuration for twenty-four dimensions. We now know the
best way to pack spheres in one, two, three, eight and twenty-four

dimensions, but no others. Viazovska's work could be the key to unlocking further dimensions.

The Abel Prize is perhaps more akin to a Nobel Prize for Mathematics, as it is often awarded to someone for a lifetime's work. Andrew Wiles won it, for example, in 2016 for his proof of Fermat's last theorem. The Abel Prize has been awarded to at least one mathematician every year since 2003, but it was only with the twentieth award that the hegemony of men was broken, when Karen Uhlenbeck won it in 2019, for her work in gauge theory and geometric analysis. These two mathematical fields have wide-ranging applications, including underpinning the unification of two fundamental forces of nature into one theory: electromagnetism and the weak nuclear force.

Mirzakhani, Viazovska and Uhlenbeck may be complete outliers. As we've seen in this book, there have always been people who have swum against the current and managed to have incredible success in mathematics. Uhlenbeck was the first woman to deliver a plenary lecture at the International Congress of Mathematicians since Emmy Noether did so in 1932, seventy years before. Our hope is that they are merely the tip of the iceberg. Over recent decades, more women have been pursuing mathematics. Parity is a long way off in terms of graduate degrees, and even more so in academic positions, but the tide is slowly starting to turn.

Geographically speaking, mathematics is opening up too, becoming a more global pursuit than ever before. Organizations such as CIMPA (International Centre for Pure and Applied Mathematics) have been set up with the intention of promoting mathematical research in mathematics in low- and middle-income countries. As of 2022, CIMPA has set up nearly four hundred short-term intensive courses in research mathematics to help this development. The founding of the African Institute for Mathematical Sciences is also leading to greater access to postgraduate training in mathematics through its pan-African network.

Problems still to solve

Mathematics is a subject about ideas. By being inclusive and bringing in talent from many diverse backgrounds, there will be more breakthroughs, and more quickly. And there is still so much for mathematicians to do.

In 2000, the Clay Mathematics Institute gathered seven important mathematical problems and put targets on their backs in the form of offering $1 million in prize money to anyone who could solve them. The initiative was inspired by David Hilbert's publication of twenty-three of the most mathematically important problems in 1900. Hilbert didn't offer a prize, but his announcement focused mathematical research for much of the twentieth century. Around half of the problems have now been solved, but only one of the so-called millennium prizes has.

On the millennium prize list are important problems – solutions to them would be monumental. A solution to the P vs NP problem, for example, would rewrite our understanding of what computers can do, and a solution to the Navier–Stokes equation, an equation central to fluid mechanics, would make us maestros of the substances that make up our world. The one that has been solved is the Poincaré conjecture, which was proven in 2002 by Grigoriy Perelman from Russia. One way of viewing this conjecture is that it tells us something about the possible shapes of the universe. If spacetime is finite

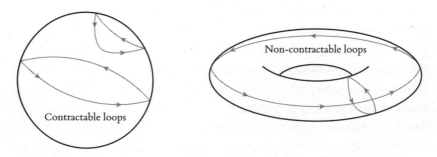

All loops on a sphere can be pulled tight without getting stuck,
but not all those on a torus.

and has the simple property that any loop you can make can be pulled tight without getting stuck, then it must be a hypersphere – a higher-dimensional version of a sphere.

Perelman was offered the $1 million prize for his proof but turned it down on the basis that he believed his contribution was equal to that of another mathematician: Richard Hamilton. Though Perelman put the nails in the conjecture's coffin, he built on Hamilton's work to such an extent that he believed Hamilton should have been equally recognized.

Other important and long-standing mathematical problems feel alluringly close to solutions but need help to get over the line. Take the twin prime conjecture. Recall, this says that there are infinitely many prime numbers that differ by just two, known as twin primes. For example, 3 and 5 are twin primes. So, too, are 5 and 7; and 11 and 13. The conjecture that there are infinitely many such primes is at least 150 years old and is a very well-known problem in number theory. It had seemed impenetrable for years, until in 2013 Yitang Zhang from China made a momentous breakthrough. He showed that there are infinitely many primes that differ by 70 million or less. This gap of 70 million is clearly a lot more than the gap of two that we're after, but until Zhang there was no boundary to the gap at all. A collaborative internet-based project, Polymath8, then sprang up to help hone Zhang's work. At the time of writing the project members have managed to reduce the gap to just 246, or 6 if a couple of other results hold true.

The future of mathematics will rely on mathematicians from around the world building on rich global mathematical traditions. It will need people from a diverse range of backgrounds to bring a diverse array of ideas and approaches. Some ideas will spring up only to be rediscovered again in a different context, while others may lie dormant for centuries. The progress of mathematics is never a straight line.

The history of mathematics has often gone hand in hand with the development of astronomy. Around the world, mathematicians looked to the sky for inspiration. Trying to comprehend how and why the celestial bodies moved helped drive the development of new techniques and ways to understand the world. This will continue throughout the twenty-first century. Technology is allowing us to see further into the depths of the universe than ever before. Telescopes sent into space will uncover ever more detail and give mathematicians even more to do. The recently online James Webb Space Telescope, for example, will be able to see the first light of the universe, as well as detect water on planets outside our solar system. The Xuntian telescope, due to launch in 2023 at the time of writing, aims to catalogue nearly 1 billion galaxies.

Long-standing problems remain to be solved, like the remaining millennium problems, but there will be new ones too. Everything from artificial intelligence, space travel, healthcare and more will require a huge amount of human ingenuity to hone and improve. The world is awash with data. As female computers once rose to the challenge of understanding an explosion of new information, so too will data scientists in our ever more data-rich society. We need new methods and techniques to understand it all. Perhaps we will need new computers too. Quantum computers that perform tasks that would take a classical computer thousands of years may soon be a reality.

Mathematics has come a long way since humanity first started scratching mathematical thoughts on to bones, clay and paper. In today's world, knowledge is more accessible than ever to more people than ever. And as mathematics and the people pursuing it become more diverse in every possible sense, so too will our means to tackle problems. We are standing on the surface of a small oblate spheroid staring into the mathematical cosmos, with twinkling theorems out there waiting to be discovered. Progress will not be linear – it never has been. And it never should be. A diversity of ideas and approaches is what makes the human pursuit of knowledge so successful. By embracing this approach, the next chapters for mathematics could be the best yet.

Acknowledgements

Kate

I can never forget that afternoon, sipping tea in a bookshop in Charing Cross, London. My long-time dream of writing a global history of mathematics would never have been realized if my agent, Max Edwards, and Tim's, Toby Mundy, had not connected us on that day. Thank you, Max and Toby, for this chance of writing a book with Tim. Thank you, Tim, for taking up this project and developing it with me. It is a great privilege to be able to work with you, and the years we spent writing together became a precious part of my life. And thank you, our editors, Connor Brown, Greg Clowes, Nick Amphlett, and Sarah Day for reading the drafts many times and helping us polish this work in such a short time.

I got lucky. People say that luck hits one when preparation meets opportunity, and my friends and colleagues are the ones who paved the way for me to seize this chance. I have been fortunate to have worked with wonderful mathematicians and historians from around the world. I owe an enormous debt to Jeremy Gray, Keith Hannabuss, Chris Hollings and Bernie Lightman for helping me develop my skills as a researcher. They have been my close friends, and I've learned a lot from them over the years. Talking to them, working with them and receiving encouragement from them allowed me to grow professionally as a historian of mathematics. Finally, I am truly grateful to my maths teacher, Bill Casselman, for his guidance and ever-lasting friendship since I was an undergraduate student at UBC.

I decided to join the Japanese space agency in the middle of writing this book. This career change impacted the book in many ways, and I thank my new colleagues and friends for giving me a new perspective.

With this book, I will travel again. I will call on my friends all over the world. I will tell them how fantastic this journey has been. The next chapter of my life will then begin. My parents, my family and my good friends, both old and new, are on board. So, what's next?

Timothy

A book takes a village, and this one is no different. However, 'acknowledgements' feels too weak a word to really capture just how important many people have been to this project. 'Immense gratitudes' would be more fitting.

I am immensely grateful to my agent, Toby Mundy, and Kate's, Max Edwards. You two encouraged us right from the beginning and guided us brilliantly through the entire process. Your sage words of wisdom helped make the book what it is and the experience what it was.

I am immensely grateful to the teams at Penguin Random House and HarperCollins. Connor Brown, you understood what we were trying to do the moment you saw the proposal, championed our idea and gave valuable feedback that helped us set our course. Greg Clowes and Nick Amphlett, you edited the first and subsequent drafts deftly and delicately, helping to better draw out what we intended to say. Sarah Day your copy-edits were invaluable. So many others also contributed behind the scenes.

I am immensely grateful to my co-author and dear friend Kate Kitagawa. Your non-stop enthusiasm, humour and expertise made working together such a pleasure. To look at what we've achieved makes me incredibly proud.

I am immensely grateful to my friends and family for providing continued love and support, especially my mother, father and sister, Big Dave and Drubs, and my Danish family. *Tusind tak til jer alle* – a thousand thanks to you all.

And I am immensely grateful to my partner, Emilie Steinmark. Without you, none of this would be possible. Emilie, *du er den eneste ene.*

Notes

Prelude

1 Newspaper article cited in Karen D. Rappaport, 'S. Kovalevsky: a mathematical lesson' in *American Mathematical Monthly* (1981), 88 (8), 564–74
2 Sónya Kovalévsky, *Her Recollections of Childhood*, translated by Isabel F. Hapgood (New York: Century, 1895), p. 316

1. In the Beginning

1 The Rhind Papyrus, vol. 1, British Museum; https://upload.wikimedia. org/wikipedia/commons/7/7b/The_Rhind_Mathematical_Papyrus,_ Volume_I.pdf

2. The Turtle and the Emperor

1 Jane Qiu, 'Ancient times table hidden in Chinese bamboo strips', *Nature*, 7 Jan. 2014; doi: 10.1038/nature.2014.14482
2 https://leibniz-bouvet.swarthmore.edu/letters/letter-j-18-may-1703-leibniz-to-bouvet/

3. A Town Called Alex

1 Plato, *The Republic*, Book VII
2 http://classics.mit.edu/Aristotle/physics.4.iv.html
3 Leonard C. Bruno and Lawrence W. Baker, *Math and Mathematicians: The History of Math Discoveries around the World* (Detroit: UXL, 1999), vol. 1, pp. 125–6

4 Pappus, Collection 3. 1: 1–8

5 K. Wider, 'Women philosophers in the ancient Greek world: donning the mantle', *Hypatia* (1986), 1 (1), 21–62; doi: 10.1111/j.1527-2001.1986. tb00521.x

6 Benjamin Wardhaugh, *The Book of Wonders: the Many Lives of Euclid's Elements* (London: William Collins, 2020), p. 34

7 Edward Watts, *Hypatia: The Life and Legend of an Ancient Philosopher* (Oxford: Oxford University Press, 2017), p. 29

8 Michael A. B. Deakin, 'Hypatia and her mathematics', *American Mathematical Monthly* (1994), 101 (3), 234–43, p. 239

9 Michael A. B. Deakin, *Hypatia of Alexandria: Mathematician and Martyr* (Amherst, NY: Prometheus Books, 2007), pp. 92–3, 97

10 Ibid., p. 97

11 *The Letters of Synesius of Cyrene*, quoted in Michael Bradley, *The Birth of Mathematics: Ancient Times to 1300* (New York: Chelsea House, 2007), p. 63

12 John of Nikiû, *Chronicle* 84: 88, quoted in Watts, *Hypatia*, p. 157, no. 4

13 John of Nikiû, *Chronicle,* 84, 87–8, 100–103

14 Watts, *Hypatia*, p. 116

15 Watts, *Hypatia*, pp. 105–6

16 Maria Dzielska, *Hypatia of Alexandria*, translated by F. Lyra (Cambridge, MA: Harvard University Press, 1996), p. 102

17 Watts, *Hypatia*, p. 5

18 M. Von Seggem, 'Notable mathematicians: from ancient times to the present', *Gale Academic Onefile* (1988), 38 (2), 257

4. The Dawn of Time

1 Eberhard Zangger and Rita Gautschy, 'Celestial aspects of Hittite religion: an investigation of the rock sanctuary Yazılıkaya', *Journal of Skyscape Archaeology* (2019), 5 (1), 5–38; https://doi.org/10.1558/jsa.37641; https:// www.newscientist.com/article/mg24232353-600-yazilikaya-a-3000-year-old-hittite-mystery-may-finally-be-solved/

5. On the Origin(s) of Zero

1 D. J. Merritt and E. M. Brannon, 'Nothing to it: precursors to a zero concept in preschoolers', *Behavioural Processes* (2013), 93, 91–7; doi: 10.1016/j.beproc.2012.11.001

2 George Gheverghese Joseph, *The Crest of the Peacock* (Princeton: Princeton University Press, 2000), opening quote

3 D. S. Hooda and J. N. Kapur, *Āryabhata: Life and Contributions* (New Delhi: New Age International Publishers, 1996), p. 78

4 Kim Plofker et al., 'The Bakhshālī manuscript: a response to the Bodleian Library's radiocarbon dating', *History of Science in South Asia* (2017), 5 (1), 134–50; doi: 10.18732/H2XT07

5 Huylebrouch, Dirk. 'Mathematics in (central) Africa before colonization,' *Anthropologica et Praehistorica* 117 (2006): 135-62

6. The House of Wisdom

1 Al-Khalili, Jim, *The House of Wisdom: How Arabic Science Saved Ancient Knowledge and Gave Us the Renaissance*. (New York: Penguin Press, 2011), p. 132

7. The Impossible Dream

1 Letter of 16 May 1643, in Lisa Shapiro, 'Princess Elizabeth and Descartes: the union of soul and body and the practice of philosophy', *British Journal for the History of Philosophy* (1999), 7 (3), 503–20, p. 505

2 Descartes in a letter to Pollot of 21 October 1643, in Carol Pal, *Republic of Women: Rethinking the Republic of Letters in the Seventeenth Century* (Cambridge: Cambridge University Press, 2012), p. 46

8. The (First) Calculus Pioneers

1 Translated by David Pingree, in Pingree 'The logic of non-Western science: mathematical discoveries in medieval India', *Daedalus* (2003), 132 (4), 45–53, p. 49

2 Quoted in Steven Strogatz, *Infinite Powers: How Calculus Reveals the Secrets of the Universe* (New York: Mariner Books, 2020), p. 200

3 Ibid., p. 201

4 All quoted in Brian E. Blank, 'Book review: *The Calculus Wars*', *Notices of the American Mathematical Society* (2009), 56 (5), 602–10, at 607

5 Quoted in ibid., 607

6 This meeting was recorded by Lady Cowper and is cited in D.Bertoloni Meli, 'Caroline, Leibniz, and Clarke', *Journal of the History of Ideas* (1999), 60 (3), 469–86, at 474

7 Letter from Caroline to Leibniz, 24 April 1716

8 Letter from Isaac Newton to Robert Hooke, 1675; https://discover.hsp.org/Record/dc-9792/Description#tabnav

9. Newtonianism for Ladies

1 https://sourcebooks.fordham.edu/mod/newton-princ.asp

2 https://www1.grc.nasa.gov/beginners-guide-to-aeronautics/newtons-laws-of-motion/

3 Translation by Simon Singh, in Singh, 'Math's hidden woman'; https://www.pbs.org/wgbh/nova/article/sophie-germain/

10. A Grand Synthesis

1 Christopher Cullen and Catherine Jami, 'Christmas 1668 and after: how Jesuit astronomy was restored to power in Beijing', *Journal for the History of Astronomy* (2020), 51 (1), 3–50, p. 18

2 Qi Han, 'Emperor, prince and literati: role of the princes in the organization of scientific activities in early Qing period', in *Current Perspectives in the History of Science in East Asia*, edited by Yung Sik

Kim and Francesca Bray (Seoul: Seoul National University, 1999), 209–16, p. 210

3 Mei Wending, *Fangcheng lung* (*On Simultaneous Linear Equations*), 1672. Quoted and translated into English in Joseph W. Dauben and Christopher Scriba, *Writing the History of Mathematics: Its Historical Development* (Basel: Birkhäuser Verlag, 2002), p. 299

4 Qi Han, 'Astronomy, Chinese and Western: the influence of Xu Guanga's views in the early and mid-Quing, in Catherine Jami, Peter Engelfriet and Gregory Blue (eds.), *Statecraft and Intellectual Renewal in Late Ming China. The Cross-cultural Synthesis of Xu Euangki* (1562–1633) (Leicten: Brill, 2001), p. 365

5 Translated by Barbara Bennet Peterson, in Peterson (ed.), *Notable Women of China* (New York. An East Gate Book, 2000), p. 344

6 Ibid

7 Ibid., p. 345

11. The Mathematical Mermaid

1 Michèle Audin, *Remembering Sofya Kovalevskaya*, (Springer: New York, 2011), p. 167

2 Translated by Leigh Whaley, in Whaley, 'Networks, patronage and women of science during the Italian Enlightenment', *Early Modern Women* (2016), 11 (1), 188

3 Translated by Beatrice Stillman, in Sofya Kovalevskaya, *A Russian Childhood* (New York: Springer-Verlag, 2013), p. 122

4 Ibid., p. 215

5 Ibid., p. 218

6 Translated by Simon Singh, in Singh, *Fermat's Enigma: The Epic Quest to Solve the World's Greatest Mathematical Problem* (Toronto: Penguin, 1998), p. 62

7 Ibid., p. 107

8 Ibid, p. 107

9 Translated by Stillman, in Kovalevskaya, *A Russian Childhood*, p. 241

10 https://mathshistory.st-andrews.ac.uk/Projects/Ellison/chapter-17/

11 Sofia to Aleksander, letter dated December 1883, in Ann Hibner Koblitz, *A Convergence of Lives: Sofia Kovalevskaia: Scientist, Writer, Revolutionary* (New York: Dover, 1993), p. 179

12 Roger Cooke, *The Mathematics of Sonya Kovalevskaya* (New York: Springer-Verlag, 1984), p. 103

13 https://mathshistory.st-andrews.ac.uk/Projects/Ellison/chapter-17/

14 Steven G. Krantz, *Mathematical Apocrypha: Stories and Anecdotes of Mathematicians and the Mathematical* (Washington DC: Mathematical Association of America, 2002), pp. 124–5. The name Sonja is used for Sophie

15 Letter from Weierstrass to Kovalevskaya, 21 September 1874, in Eva Kaufholz-Soldat, ' "[. . .] the first handsome mathematical lady I've ever seen!": On the role of beauty in portrayals of Sofia Kovalevskaya', *Journal of the British Society for the History of Mathematics* (2017), 32 (3) 198–213, at 209

16 E. T. Bell, *Men of Mathematics,* vol. 2 (London: Penguin, 1953), p. 468

17 Kaufholz-Soldat, '[. . .] the first handsome mathematical lady I've ever seen!', 209, 211

18 Sónya Kovalévsky, *Her Recollections of Childhood*, translated by Isabel F. Hapgood (New York: The Century Co., 1859), p. 316

12. Revolutions

1 W. K. Bühler, *Gauss: A Biographical Study* (Berlin: Springer-Verlag, 1981), p. 106

2 Carl B. Boyer, *A History of Mathematics* (Princeton: Princeton University Press, 1985), p. 587

3 June Barrow-Green, Jeremy Gray and Robin Wilson, *The History of Mathematics: A Source-Based Approach*, vol. 2 (Providence: MAA Press, 2022), p. 394

4 Ibid., p. 395

5 https://mathshistory.st-andrews.ac.uk/OfTheDay/oftheday-11-08/

6 https://www.nytimes.com/2012/03/27/science/emmy-noether-the-most-significant-mathematician-youve-never-heard-of.html

7 *Dokumente zu Emmy Noether* (n. d.). Compiled by Peter J. Raquette., 1.2,
   9, from Helmut Hasse to the Curator of the University of Göttingen,
   https://www.mathi.uni-heidelberg.de/.quette/Translenptioner/
   DOKNOE_070228.pdf
8 Louise S. Grinstein and Paul J. Campbell, 'Anna Johnson Pell Wheeler:
   her life and work', *Historia Mathematica* (1982), 9 (1), 37–53, at 42
9 Written on 3 May 1935, and published on 5 May 1935

13. =

1 https://www.whitehousehistory.org/benjamin-bannaker
2 Ibid
3 *Washington Post*, 2 Dec. 1969
4 Susan E. Kelly, Carly Shinners and Katherine Zoroufy. 'Euphemia
   Lofton Haynes: bringing education closer to the "goal of perfection" ',
   Notices of the AMS (Oct. 2017), 64 (9), 995–1002, at 997
5 Ibid., 1000
6 Donald J. Albers and G. L. Alexanderson, *Mathematical People: Profiles
   and Interviews*, (Wellesley, MA: A. K. Peters, 2nd edn 2008) p. 20
7 Ibid. p. 19
8 https://stat.illinois.edu/news/2020-07-17/david-h-blackwell-profile-
   inspiration-and-perseverance
9 Morris H. DeGroot, 'A conversation with David Blackwell', *Statistical
   Science* (1986), 1 (1), 40–53, at 41

14. Mapping the Stars

1 https://vcencyclopedia.vassar.edu/distinguished-alumni/
   antonia-maury/
2 https://makerswomen.tumblr.com/post/171799965773/they-were-going-
   to-the-moon-i-computed-the-path

15. Number-crunching

1 G. H. Hardy, 'The Indian mathematician Ramanujan', *American Mathematical Monthly* (1937), 44 (3), 137–55, at 152

2 Harald August Bohr, *Collected Mathematical Works* (Copenhagen: Dansk matematisk forening, 1952), p. xxvii

3 Letter of Ramanujan, 16 January 1913

4 G. H. Hardy, Obituary of S. Ramanujan, *Nature* (1920), 105, 494–5; https://doi.org/10.1038/105494a0

5 Robert Kanigel, *The Man Who Knew Infinity: A Life of the Genius Ramanujan* (New York: Washington Square Press, 1991), p. 167

6 Shawnee L. McMurran and James J. Tattersall, 'The mathematical collaboration of M. L. Cartwright and J. E. Littlewood', *American Mathematical Monthly* (1996), 103 (10), 833–45, p. 836

7 W. K. Hayman, 'Dame Mary (Lucy) Cartwright, D.B.E., 17 December 1900–3 April 1998', *Biographical Memoirs of Fellows of the Royal Society* (2000), 46, 19–35, at 31; https://doi.org/10.1098/rsbm.1999.0070

8 https://www.bbc.com/news/magazine-21713163

9 G. H. Hardy, *A Mathematician's Apology* (Cambridge: Cambridge University Press, 1940)

Epilogue

1 https://link.springer.com/article/10.1007/s43545-021-00098-6

2 https://math.mit.edu/wim/2019/03/10/national-mathematics-survey/

3 https://op.europa.eu/en/web/eu-law-and-publications/publication-detail/-/publication/67d5a207-4da1-11ec-91ac-01aa75ed71a1

4 https://journals.plos.org/plosone/article?id=10.1371/journal.pone.0241915

5 https://www.livescience.com/breakthrough-prize-mathematics-2019-winners.html

Further Readings

1. In the Beginning

Barrow-Green, June, Jeremy Gray and Robin Wilson, *The History of Mathematics: A Source-Based Approach*, vols. 1 and 2 (Providence: The Mathematical Association of America, 2019, 2022)

Bradley, Michael, *The Birth of Mathematics: Ancient Times to 1300* (New York: Chelsea House, 2007)

Bruno, Leonard C., and Lawrence W. Baker, *Math and Mathematicians: The History of Math Discoveries around the World* (Detroit: UXL, 1999)

2. The Turtle and the Emperor

Cullen, Christopher, *Astronomy and Mathematics in Ancient China: The Zou bi suan jing* (Cambridge: Cambridge University Press, 1996)

Cullen, Christopher, *Heavenly Numbers: Astronomy and Authority in Early Imperial China* (Oxford: Oxford University Press, 2017)

Dauben, Joseph W., 'Suan Shu Shu: a book on numbers and computations', *Archive for History of Exact Sciences* (2008), 62, 91–178

Lam Lay Yong and Ang Tian Se, *Fleeting Footsteps: Tracing the Conception of Arithmetic and Algebra in Ancient China* (rev. edn) (River Edge, NJ: World Scientific, 2004)

Martzloff, Jean-Claude, *A History of Chinese Mathematics* (Berlin: Springer, 1987)

Swann, Nancy Lee, *Pan Chao, Foremost Woman Scholar of China, First Century AD. Background, Ancestry, Life, and Writings of the Most Celebrated Chinese Woman of Letters* (New York: Century Co., c.1932)

3. A Town Called Alex

Deakin, Michael A. B., 'Hypatia and her mathematics', *American Mathematical Monthly* (1994), 101 (3), 234–43

Deakin, Michael A. B., *Hypatia of Alexandria: Mathematician and Martyr* (Amherst, NY: Prometheus Books, 2007)

Dzielska, Maria, *Hypatia of Alexandria*, translated by F. Lyra (Cambridge, MA: Harvard University Press, 1996)

Knorr, Wilbur Richard, *Textual Studies in Ancient and Medieval Geometry* (Boston: Birkhäuser, 1989)

Lawrence, Snezana and Mark McCartney (eds.), *Mathematicians and Their Gods: Interactions between Mathematics and Religious Beliefs* (Oxford: Oxford University Press, 2015)

McLaughlin, Gráinne, 'The logistics of gender from classical philosophy', in *Women's Influence on Classical Civilization*, edited by Fiona McHardy and Eireann Marshall (London: Routledge, 2004), 7–25

Wardhaugh, Benjamin, *The Book of Wonders: The Many Lives of Euclid's Elements* (London: William Collins, 2020)

Watts, Edward, *Hypatia: The Life and Legend of an Ancient Philosopher* (Oxford: Oxford University Press, 2017)

4. The Dawn of Time

Denny, Mark, *Ingenium: Five Machines that Changed the World* (Baltimore: Johns Hopkins University Press, 2007)

Gleick, James, *Time Travel: A History* (London: HarperCollins, 2016)

Hawking, Stephen, *A Brief History of Time* (London: Bantam Press, 1988)

Lebrun, David, *Cracking the Maya Code*, history documentary (2008), pbs.org/wgbh/nova/mayacode

North, John, *God's Clockmaker: Richard of Wallingford and the Invention of Time* (London: Continuum, 2005)

Ogle, Vanessa, *The Global Transformation of Time, 1870–1950* (Cambridge, MA: Harvard University Press, 2015)

Robson, Eleanor, 'The tablet house: a scribal school in old Babylonian Nippur', *Revue d'Assyriologie et d'Archéologie Orientale* (2001), 93, 39–66

Rovelli, Carlo, *The Order of Time* (London: Penguin, 2018)

5. On the Origin(s) of Zero

Brown, Nancy Marie, *The Abacus and the Cross: The Story of the Pope Who Brought the Light of Science to the Dark Ages* (New York: Basic Books, 2010)

Burnett, Charles, *Numerals and Arithmetic in the Middle Ages* (Farnham: Ashgate Variorum, 2010)

Clark, Walter Eugene (trans. and ed.), *The Āryabhaṭiya of Āryabhaṭa: An Ancient Indian Work on Mathematics and Astronomy* (Chicago: University of Chicago Press, 1930)

Cooke, Roger, *The History of Mathematics: A Brief Course* (New York: Wiley, 1997)

Dutta, Amartya Kumar, 'Āryabhaṭa and axial rotation of Earth: 3. a brief history', *Resonance* (2006), 58–72

Eraly, Abraham, *The First Spring: The Golden Age of India* (New Delhi: India Viking, 2011)

Hammer, Joshua, *The Bad-Ass Librarians of Timbuktu: And Their Race to Save the World's Most Precious Manuscripts* (New York: Simon and Schuster, 2016)

Joseph, George Gheverghese, *The Crest of the Peacock: Non-European Roots of Mathemetics* (Princeton: Princeton University Press, 2000)

Padmanabhan, Thanu (ed.), *Astronomy in India: A Historical Perspective* (New Delhi: Indian National Science Academy and Springer India, 2014)

Saad, Elias N., *Social History of Timbuktu: The Role of Muslim Scholars and Notables 1400–1900* (Cambridge: Cambridge University Press, 1983)

6. The House of Wisdom

Al-Khalili, Jim, *The House of Wisdom: How Arabic Science Saved Ancient Knowledge and Gave Us the Renaissance* (New York: Penguin Press, 2011)

Brooks, Michael, 'Mathematics in Africa has been written out of history books – it's time we reminded the world of its rich past', *Independent*, 24 Oct. 2021; https://www.independent.co.uk/voices/african-mathematics-black-history-b1944288.html

Burnett, Charles, *The Introduction of Arabic Learning into England* (London: British Library, 1997)

Knuth, Donald E., 'Algorithms in modern mathematics and computer science', in *Algorithms in Modern Mathematics and Computer Science*, edited by A. P. Ershov and D. E. Knuth (Berlin: Springer, 1981), pp. 82–99

Loop, Jan, Alastair Hamilton and Charles Burnett (eds.), *The Teaching and Learning of Arabic in Early Modern Europe* (Leiden: Brill, 2017)

Lyons, Jonathan, *The House of Wisdom: How the Arabs Transformed Western Civilization* (New York: Bloomsbury, 2009)

Roberts, Victor, 'The planetary theory of Ibn al-Shāṭir: latitudes of the planets', *Isis* (1996), 57 (2), 208–19

Saliba, George, *Islamic Science and the Making of the European Renaissance* (Cambridge MA: MIT Press, 2007)

Zemanek, Heinz, 'Al-Khorezmi: his background, his personality, his work, and his influence', in *Algorithms in Modern Mathematics and Computer Science*, edited by A. P. Ershov and D. E. Knuth (Berlin: Springer-Verlag, 1981), pp. 1–81

7. The Impossible Dream

The Correspondence between Princess Elisabeth of Bohemia and René Descartes, edited and translated by Lisa Shapiro (Chicago: University of Chicago Press, 2007)

Devlin, Keith J., *The Unfinished Game: Pascal, Fermat, and the Seventeenth-Century Letter that Made the World Modern* (New York: Basic Books, 2008)

Gorroochurn, Prakash, 'Thirteen correct solutions to the "Problem of Points" and their histories', *Mathematical Intelligencer* (2014), 36, 56–64, doi: 10.1007/s00283-014-9461-5

Kitagawa, Tomoko L., 'Passionate souls: Elisabeth of Bohemia and René Descartes', *Mathematical Gazette* (2021), 105 (563), 193–200

Pal, Carol, *Republic of Women: Rethinking the Republic of Letters in the Seventeenth Century* (Cambridge: Cambridge University Press, 2012)

Raj, Kapil, *Relocating Modern Science: Circulation and the Construction of Knowledge in South Asia and Europe, 1650–1900* (New York: Palgrave Macmillan, 2007)

Remmert, Volker R., 'Inventing tradition in 16th- and 17th-century mathematical sciences: Abraham as teacher of arithmetic and astronomy', *Mathematical Intelligences* (2015), 37 (3), 55–9

Riskin, Jessica, 'Machines in the garden', *Republics of Letters: A Journal for the Study of Knowledge, Politics, and the Arts* (2010), 1 (2), 16–43

Schiebinger, Londa, *The Mind Has No Sex?: Women in the Origins of Modern Science* (Cambridge, MA: Harvard University Press, 1989)

Shapiro, Lisa, 'Princess Elizabeth and Descartes: The union of soul and body and the practice of philosophy', *British Journal for the History of Philosophy* (1999), 7 (3), 503–20

8. The (First) Calculus Pioneers

Bertoloni Meli, D., 'Caroline, Leibniz, and Clarke', *Journal of the History of Ideas* (1999), 60 (3), 469–86

The Birth of Calculus, BBC documentary (1986)

Blank, Brian E., 'Book review: *The Calculus Wars*', *Notices of the American Mathematical Society* (2009), 56 (5), 602–10

Fara, Patricia, *Life after Gravity: Isaac Newton's London Career* (Oxford: Oxford University Press, 2021)

Hellman, Hal, *Great Feuds in Mathematics: Ten of the Liveliest Disputes Ever* (Hoboken, NJ: John Wiley and Sons, 2006)

Joseph, George Gheverghese, *A Passage to Infinity: Medieval Indian Mathematics from Kerala and Its Impact* (New Delhi: Sage, 2009)

Katz, Victor J., 'Ideas of calculus in Islam and India', *Mathematics Magazine* (1995), 68 (3), 163–74

Pingree, David, 'The logic of non-Western science: mathematical discoveries in medieval India', *Daedalus* (2003), 132 (4), 45–53.

Rajagopal, C. T. and Rangachari, M., S., 'On an untapped source of medieval Keralese mathematics', *Archive for History of Exact Sciences* (1978), 18 (2), 89–102

Rajagopal, C. T. and Venkataraman, A., 'The sine and cosine power-series in Hindu mathematics', *Journal of the Royal Asiatic Society of Bengal – Science* (1949), 15, 1–13

Sarma, K. V., *A History of the Kerala School of Hindu Astronomy (in Perspective)* (Hoshiarpur: Vishveshvaranand Institute, 1972)

Stillwell, John, *Mathematics and Its History* (New York: Springer, 2010)

Strogatz, Steven, *Infinite Powers: How Calculus Reveals the Secrets of the Universe* (New York: Mariner Books, 2020)

9. Newtonianism for Ladies

Arianrhod, Robyn, *Seduced by Logic: Émilie du Châtelet, Mary Somerville and the Newtonian Revolution* (New York: Oxford University Press, 2012)

Boran, Elizabethanne and Mordechai Feingold (eds.), *Reading Newton in Early Modern Europe* (Leiden: Brill, 2017)

Brasch, Frederick E., 'The Newtonian epoch in the American colonies (1680–1783)', *Proceedings of the American Antiquarian Society* (1939), 49 (2), 314–32

Cifarelli, Luisa and Raffaella Simili (eds.), *Laura Bassi – The World's First Woman Professor in Natural Philosophy: An Iconic Physicist in Enlightenment Italy* (Cham: Springer Nature, 2020)

Ferreiro, Larrie D., *Measure of the Earth: The Enlightenment Expedition that Reshaped Our World* (New York: Basic Books, 2013)

Findlen, Paula, 'Calculations of faith: mathematics, philosophy, and sanctity in 18th-century Italy (new work on Maria Gaetana Agnesi)', *Historia Mathematica* (2011), 38 (2), 248–91

Matytsin, Anton M., *The Specter of Skepticism in the Age of Enlightenment* (Baltimore: Johns Hopkins University Press, 2016)

Mazzotti, Massimo, *The World of Maria Gaetana Agnesi, Mathematician of God* (Baltimore: Johns Hopkins University Press, 2018)

Mazzotti, Massimo, 'Newton for ladies: gentility, gender, and radical culture', *British Journal for the History of Science* (2004), 37 (2), 119–46

Terrall, Mary, *The Man Who Flattened the Earth: Maupertuis and the Sciences in the Enlightenment* (Chicago: Chicago University Press, 2002)

Zinsser, Judith P., *Emilie du Châtelet: Daring Genius of the Enlightenment* (New York: Penguin, 2007)

10. A Grand Synthesis

Cullen, Christopher and Catherine Jami, 'Christmas 1668 and after: how Jesuit astronomy was restored to power in Beijing', *Journal for the History of Astronomy* (2020), 51 (1), 3–50

Dauben, Joseph W. and Christopher Scriba (ed.), *Writing the History of Mathematics: Its Historical Development* (Basel: Birkhäuser Verlag, 2002)

Elman, Benjamin, *On Their Own Terms: Science in China, 1550–1900* (Cambridge, MA: Harvard University Press, 2005)

Engelfriet, Peter M., *Euclid in China: The Genesis of the First Chinese Translation of Euclid's Elements Books I–VI Jihe Yuanben; Beijing, 1607) and Its Reception up to 1723* (Leiden: Brill, 1998)

Gerritsen, Anne and Giorgio Riello (eds.), *The Global Lives of Things: The Material Culture of Connections in the Early Modern World* (Abingdon: Routledge, 2016)

Han Qi, 'Emperor, prince and literati: role of the princes in the organization of scientific activities in early Qing period', in *Current Perspectives in the History of Science in East Asia*, edited by Yung Sik Kim and Francesca Bray (Seoul: Seoul National University, 1999), 209–16.

Ho, Clara Wing-chung (ed.), *Biographical Dictionary of Chinese Women: The Qing Period, 1644–1911* (Armark, New York: M. E. Sharpe, 1998)

Jami, Catherine, *The Emperor's New Mathematics: Western Learning and Imperial Authority during the Kangxi Reign (1662–1722)* (Oxford: Oxford University Press, 2012)

Jami, Catherine, 'Revisiting the Calendar Case (1664–1669): science, religion, and politics in early Qing Beijing', *Korean Journal for the History of Science* (2015), 37 (2), 459–77

Lam Lay Yong and Shen Kangshen, 'Methods of solving linear equations in traditional China', *Historia Mathematica* (1989), 16 (2), 107–22

Lü Lingfeng, 'Eclipses and the victory of European astronomy in China', *East Asian Science, Technology, and Medicine* (2007), 27, 127–45

Mungello, D. E., *The Great Encounter of China and the West, 1500–1800*, 4th edn (Lanham, MD: Rowman & Littlefield, 2012)

Peterson, Barbara Bennett (ed.), *Notable Women of China: Shang Dynasty to the Early Twentieth Century* Armonk (New York: M. E. Sharpe, 2000)

Subrahmanyam, Sanjay, *Europe's India: Words, People, Empires, 1500–1800* (Cambridge, MA: Harvard University Press, 2017)

Lǐ Yan and Dù Shíràn, *Chinese Mathematics: A Concise History*, translated by John N. Crossley and Anthony W. C. Lun (Oxford: Clarendon Press, 1987)

11. The Mathematical Mermaid

Audin, Michèle, *Remembering Sofya Kovalevskaya* (London: Springer, 2011)

Cooke, Roger, *The Mathematics of Sonya Kovalevskaya* (New York: Springer-Verlag, 1984)

Kaufholz-Soldat, Eva, ' "[. . .] the first handsome mathematical lady I've ever seen!": on the role of beauty in portrayals of Sofia Kovalevskaya', *Journal of the British Society for the History of Mathematics* (2017), 32 (3), 198–213

Koblitz, Ann Hibner, *A Convergence of Lives: Sofia Kovalevskaia: Scientist, Writer, Revolutionary* (New Brunswick, NJ. Rutgers University Press, 1993)

Kovalevskaya, Sofya, *A Russian Childhood*, translated by Beatrice Stillman (New York: Springer, 2013)

Laubenbacher, Reinhard and David Pengelley, ' "Voici ce que j'ai trouvé": Sophie Germain's grand plan to prove Fermat's Last Theorem', *Historica Mathematica* (2010), 37 (4), 641–92

Munro, Alice, *Too Much Happiness* (Toronto: McClelland & Stewart, 2009)

Musielak, Dora, *Sophie Germain: Revolutionary Mathematician*, 2nd edn (Cham: Springer, 2020)

Van Tiggelen, Brigitte, 'Emilie du Châtelet and the nature of fire: *Dissertation sur la nature et la propagation du feu*', in Annette Lykknes and Brigitte Van Tiggelen (eds.), *Women in Their Element: Selected Women's Contributions to the Periodic System* (Hackersack, NJ: World Scientific, 2019), pp. 70–84

Singh, Simon, *Fermat's Enigma: The Quest to Solve the World's Greatest Mathematical Problem* (Toronto: Penguin, 1998)

12. Revolutions

Boyer, C. B., *A History of Mathematics* (Princeton: Princeton University Press, 1985)

Brading, Katherine, 'A note on general relativity, energy conservation, and Noether's theorems', in *The Universe of General Relativity*, edited by A. J. Kox and Jean Eisenstaedt (Boston: Birkhäuser, 2005), 125–35

Brylevskaya, Larisa I., 'Lobachevsky's geometry and research of geometry of the universe', *Publications of the Astronomical Observatory of Belgrade* (2008), 85, 129–34

Gray, Jeremy, 'Gauss and non-Euclidean geometry', in *Mathematics and Its Applications: Janús Bolyai Memorial Volume,* vol. 581, edited by András Prékopa and Emil Molnár (New York: Springer, 2006), pp. 61–80

Kitagawa, Tomoko L., 'Moscow, Oxford, or Princeton: Emmy Noether's move from Göttingen (1933)', in *The Philosophy and Physics of Noether's Theorems: A Centenary Volume*, edited by James Read and Nicholas J. Teh (Cambridge: Cambridge University Press, 2022), pp. 52–65

Tobies, Renate, *Felix Klein: Visions for Mathematics, Applications, and Education*, translated by Valentine A. Pakis (Cham: Birkhäuser, 2021)

Roselló, Joan, *Hilbert, Göttingen and the Development of Modern Mathematics* (Newcastle upon Tyne: Cambridge Scholars Publishing, 2019)

Rowe, David E. and Mechthild Koreuber, *Proving It Her Way: Emmy Noether, a Life in Mathematics* (Cham: Springer, 2020)

History Working Group [at the Institute for Advanced Study], 'Emmy Noether's paradise: how IAS helped support the first female professor in Germany when she became a displaced refugee', *The Institute Letter Spring 2017*

Inside Einstein's Mind: The Enigma of Space and Time, BBC documentary (2015), writer/director Jamie E. Lochhead.

13. =

Black, Robert, *David Blackwell and the Deadliest Duel* (Unionville, NY: Royal Fireworks Press, 2019)

Blackwell, David, Leo Breiman and A. J. Thomasian, 'Proof of Shannon's transmission theorem for finite-state indecomposable channels', *Annals of Mathematical Statistics* (1958), 29 (4), 1209–20

Cerami, Charles, *Benjamin Bannaker: Surveyor, Astronomer, Publisher, Patriot* (New York: J. Wiley & Sons, 2002)

Cox, Elbert, 'On a class of interpolation functions for system of grading', *Journal of Experimental Education* (1947), 15 (4), 331–41

DeGroot, Morris H., 'A conversation with David Blackwell', *Statistical Science* (1986), 1 (1), 40–53

Donaldson, James A. and Richard J. Fleming, 'Elbert F. Cox: an early pioneer', *American Mathematical Monthly* (2000), 107 (2), 105–28

Interview: 'David Blackwell: Working at Howard University', https://www.youtube.com/watch?v=sMzntPFemmM&t=367s

Kelly, Susan E., Carly Shinners and Katherine Zoroufy, 'Euphemia Lofton Haynes: bringing education closer to the "goal of perfection"', *Notices of AMS* (Oct. 2017), 995–1003

Shannon, C. E., 'A mathematical theory of communication', *Bell System Technical Journal* (1948), 27, 379–423, 623–56

Slater, Robert Bruce, 'The blacks who first entered the world of white higher education', *Journal of Blacks in Higher Education* (1994), 4, 47–56

14. Mapping the Stars

Abbate, Janet, *Recoding Gender: Women's Changing Participation in Computing* (Cambridge, MA: MIT Press, 2012)

Dick, Steven J. (ed.), *Remembering the Space Age* (Washington DC: National Aeronautics and Space Administration Office of External Relations History Division, 2008)

Geiling, Natasha, 'The women who mapped the universe and still couldn't get any respect', *Smithsonian Magazine* (18 Sept. 2013); https://www.smithsonianmag.com/history/the-women-who-mapped-the-universe-and-still-couldnt-get-any-respect-9287444/

Glass, I. S., *The Royal Observatory at the Cape of Good Hope: History and Heritage* (Cape Town: Mons Mensa Publishing, 2015)

Haley, Paul A., 'Entente céleste: David Gill, Ernest Mouchez, and the Cape and Paris Observatories, 1878–92', *Antiquarian Astronomer* (2016), 10, 13–37

Hearnshaw, J. B., *The Measurement of Starlight: Two Centuries of Astronomical Photometry* (Cambridge: Cambridge University Press, 1996)

Johnson, George, *Miss Leavitt's Stars: The Untold Story of the Woman Who Discovered How to Measure the Universe* (New York: W. W. Norton, 2005)

Jones, Derek, 'The scientific value of the Carte du Ciel', *Astronomy & Geo-physics* (2000), 41(5), 5.16–5.20

Nakamura, Tsuko and Wayne Orchiston (eds.), *The Emergence of Astro-physics in Asia: Opening a New Window on the Universe* (Cham: Springer, 2017)

Schuster, William J. and Marco Arturo Moreno-Corral, 'Astronomy in Mex-ico during the first years of the IAU', in *Under One Sky: The IAU Centenary Symposium*, edited by Christiaan Sterken, John Hearnshaw and David Valls-Gabaud (Cambridge: Cambridge University Press, 2019)

Sobel, Dava, *The Glass Universe: How the Ladies of the Harvard Observatory Took the Measure of the Stars* (New York: Viking, 2016)

Spangenburg, Ray and Kit Moser, *African Americans in Science, Math, and Invention* (New York: Facts On File, 2003)

Turner, H. H., *The Great Star Map: Being a Brief General Account of the Inter-national Project Known as the Astrographic Chart* (New York: E. P. Dutton and Company, 1912)

15. Number-crunching

Cartwright, M. L., 'Mathematics and thinking mathematically', *American Mathematical Monthly* (1970), 77 (1), 20–28

Grattan-Guinness, I., 'Russell and G. H. Hardy: a study of their relation-ship', *Journal of the Bertrand Russell Archives* (1991–2), 11, 165–79

Hardy, G. H., *A Mathematician's Apology* (Cambridge: Cambridge Univer-sity Press, 1940)

Hodges, Andrew, *The Enigma* (London: Vintage, 2012)

Kanigel, Robert, *The Man Who Knew Infinity: A Life of the Genius Ramanujan* (New York: Washington Square Press, 1991)

Kitagawa, Tomoko L. and Eder Kikianty, 'A history of mathematics in South Africa: modern milestones', *Mathematical Intelligencer* (2021), 43 (4), 33–47

Littlewood, J. E., *A Mathematician's Miscellany* (London: Methuen, 1953)

McMurran, Shawnee L. and James J. Tattersall, 'The mathematical collabo-ration of M. L. Cartwright and J. E. Littlewood', *American Mathematical Monthly* (1996), 103 (10), 833–45

Ono, Ken and Robert Schneider, 'We're still untangling Ramanujan's mathematics 100 years after he died', *New Scientist*, 22 April 2020, https://www.newscientist.com/article/mg24632792-600-were-still-untangling-ramanujans-mathematics-100-years-after-he-died/

Williams, H. Paul, 'Stanley Skewes and the Skewes number', *Journal of the Royal Institution of Cornwall* (2007), 70–75; http://eprints.lse.ac.uk/31662/

Winterson, Jeanette, *12 Bytes: How We Got Here, Where We Might Go Next* (London: Jonathan Cape, 2021), https://www.bbc.co.uk/

Epilogue

Green, Judy, 'How many women mathematicians can you name?' *Math Horizons* (2001), 9 (2), 9–14

Selin, Helaine (ed.), *Mathematics across Cultures: The History of Non-Western Mathematics* (Dordrecht: Springer Science + Business Media, 2000)

Wilson, Robin, *Number Theory: A Very Short Introduction* (Oxford: Oxford University Press, 2020)

Index

Page numbers in italics indicate illustrations. Important numbers discussed in the text are cited in ascending order at the beginning of the index.

International Congress of
    Mathematicians (ICM)
    200–201
International Mathematical
    Olympiad 257
Iran 9, *10*, 89, 101, 257
Iraq 9, *10*, *90*, 101
    see also Baghdad
irrigation systems 9, 16
Ishango bone 7, 7–8
Islam 84, 91, 92
Islamic empire 89, 90, 101
Italians 86, 109, 153–4, 163–4, 221
Italy 103, 111, 151–2, 172

Jainism 74, 75
Jain(s)
    belief 75
    exponentiation 76–7
    mathematicians 74–7, 80
    number system 75–7
    people 74–6
    on tastes 76
    triangle discovery 76
    understanding of the universe 74–5
James I 112
Janssen, Jules 171
Japan 40, 157, 201–2, 206, 212, 254
Japanese, the 40, 128, 201, 202, 212,
    254, 263
Jesuit(s) *160*
    astronomers 156, 159–61, 164
    mathematicians 157, 164–5, 167, 168
    missionaries 29, 67, 159–61, *163*,
        163–4, 169–70
Jews 53, 89, 92, 204, 207
Jim Crow laws 211
Jin, Guliang *39*
John of Nikiû 53, 54
Johnson, Katherine 231–3
Jones, Alexander 47
Jones, Katherine 117
Jones, William 32
Joseph, George Gheverghese 138

*Journal of the Indian Mathematical
    Society* 238
Jump, Mary 226
Juno 221, *222*
Jupiter 58–9, 100, 208, 221–2
Jyesthadeva 123

Kangxi (Emperor) 29, 67, 158, 161,
    162–4, *163*, 167
Kaufholz-Soldat, Eva 188
Kazan 196
Kazan Messenger, The 197
Keill, John 135
Kepler, Johannes 51, 140–42
Kerala 119–23, 134, 137–8
    Calicut *119*
    Cochin 120
    Sangamagrama 4, 122
Khan, Hulegu 101
Khayyam, Omar 193
Khmer empire 83
Khwarazm 93
Kidinnu 59
Kingsley, Charles 55
Klein bottle 199–200, *201*
Klein, Felix 187, 199–201, 203, 204
Korea 40, 157, 202
Korvin-Krukovskaya, Sophie *see*
    Kowalevski, Sophie
Kovalevsky, Vladimir 176–7,
    181–2, 183
Kowalevski, Sophie 174–9, *177*, 181–9,
    203, 227
    daughter, Foufie 181–2, 183
    death of 187
    distortion of legacy of 5, 187–8
    doctorate 178
    early life 4, 174–5
    education 4–5, 175–7
    father, Vasily 4, 174–7
    as first female professor 5, 171, 183
    husband of 176–7, 181–2, 183
    the 'mathematical mermaid' 182,
        184–6